Power Systems

Electrical power has been the technological foundation of industrial societies for many years. Although the systems designed to provide and apply electrical energy have reached a high degree of maturity, unforeseen problems are constantly encountered, necessitating the design of more efficient and reliable systems based on novel technologies. The book series Power Systems is aimed at providing detailed, accurate and sound technical information about these new developments in electrical power engineering. It includes topics on power generation, storage and transmission as well as electrical machines. The monographs and advanced textbooks in this series address researchers, lecturers, industrial engineers and senior students in electrical engineering.

More information about this series at http://www.springer.com/series/4622

Anna Hnydiuk-Stefan

Dual-Fuel Gas-Steam Power Block Analysis

Methodology and Continuous-Time Mathematical Models

Anna Hnydiuk-Stefan
Opole, Poland

ISSN 1612-1287 ISSN 1860-4676 (electronic)
Power Systems
ISBN 978-3-030-13215-6 ISBN 978-3-030-03050-6 (eBook)
https://doi.org/10.1007/978-3-030-03050-6

Softcover re-print of the Hardcover 1st edition 2019

This Springer imprint is published by the registered company Springer Nature Switzerland AG
The registered company address is: Gewerbestrasse 11, 6330 Cham, Switzerland

Contents

Notations

A	Depreciation rate
a_{el}, a_{coal}, b_{CO_2}	Control variables for prices of electricity, fuel, CO_2 emission allowances
a_{CO_2}, a_{CO}, a_{SO_2}, a_{NO_X}, a_{dust}	Control variables for specific tariffs for emission of pollutants into the environment CO_2, CO, SO_2, NO_x, dust
a_{CHP}	Control (exponent $e_{CHP}^{t=0}e^{a_{CHP}t}$) change in price of the certificate of origin of electricity produced in high-efficiency cogeneration
ARR	Accounting rate of return
b	Duration of the investment
CF	Cash flow
D	Annual operation net profit of the power station or heat and power plant
$DPBP$	Discounted payback period
e^{-rt}	Money discounting factor
e_h, e_{el}, e_{coal}	Specific cost of heat, electricity, and coal
$E_{CHP,\,A}$	Annual amount of electricity produced in high-efficiency cogeneration in accordance with Directive 2004/8/EC of the European Parliament
$E_{el,\,A}$	Net annual electricity output from a heat and power plant or a power station
$E_{ch,\,A}$	Annual use of the chemical energy of the fuel heat and power plant or a power station
E_{ch}^{gas}	Chemical energy of gas combustion in the gas turbine
E_{ch}^{coal}	Chemical energy of coal combustion in the boiler
E_{el}^{GT}	Gross electrical output of the gas turbogenerator
E_{el}^{ST}	Gross electrical output of the steam turbogenerator

F	Time variable interest (financial cost) relative to the value of investment
i	Specific investment (per unit of power)
IPP	Independent power producer
IRR	Internal rate of return
J	Investment expenditure
k_h	Specific cost of heat production in a heat and power plant
$k_{h,av}$	Average specific cost of heat production
k_{el}	Specific cost of electricity production in a power unit
K_e	Annual exploitation cost of heat and power plant and power station
K_{coal}	Cost of fuel
K_{sw}	Cost of supplementing water in system circulation
K_{sal}	Cost of remuneration including overheads
K_{serv}	Cost of maintenance and overhaul
K_m	Cost of non-energy resources and supplementary materials
K_{env}	Cost associated with the use of the environment (including: charges for emission of flue gases into the atmosphere, waste disposal, waste storage)
K_{cap}	Capital cost
Kt	Taxes, local charges, and insurance
K_{CO_2}	Cost of purchasing CO_2 emission allowances
N	Calculated exploitation period of the installation
N_{el}	Gross electrical capacity of a gas or steam turbogenerator
NPV	Net present value
p	Tax rate on profit before tax
P	Income tax on gross profit
Q_A	Annual net heat output of a combined heat and power unit
Q_A	Annual net production of heat, GJ/a
Q_{con}	Heat from steam condensing in the condenser of the steam turbine
q_{par}	Ratio of chemical energy of gas in relation to the chemical energy of coal in the parallel system
q_{ser}	Ratio of chemical energy of gas in proportion to the chemical energy of coal in the series system
R	Loan installment
r	Discount rate
ROE	Return on equity
ROI	Return on investment

S_A	Annual revenues from the operation of heat and power station or from power plant
$SPBP$	Simple payback period
t	Time
T	Exploitation period of the installation calculated in years
u	Ratio of chemical energy of fuel to its total use, for which it is not necessary to purchase CO_2 emission allowances
v_m	Relative value of heat and electricity market
$x_{sw,m,was}$	Coefficient accounting for the cost of supplementing water, auxiliary materials, waste disposal, slag storage, and waste
$x_{sal,t,ins}$	Coefficient accounting for the cost of remuneration, taxes, insurance
z	Ratio of freezing an investment
Z	Annual operation gross profit of the power station or heat and power plant
δ_{serv}	Rate of fixed costs relative to investment (cost of maintenance and overhaul)
ε_{el}	Internal load of power station or heat and power plant
η	Efficiency
ρ	Emission of harmful combustion products into the atmosphere
ρ	Rate of depreciation including interest
η_B	Gross boiler efficiency
η_c	Energy efficiency of heat and electricity production
η_{el}	Gross efficiency of power plant
η_{HRSG}	Gross efficiency of the heat recovery steam generator
η_{SH}	Energy efficiency of steam header used to feed steam into the turbine
η_{GT}	Gross energy efficiency of the gas turbine
η_{ST}	Energy efficiency of steam turbine
η_{me}	Electromechanical efficiency of the steam turbogenerator
σ_A	Annual cogeneration index

Chapter 1
Introduction

For the things of this world cannot be made known without a knowledge of mathematics.

Roger Bacon (1214–1294)

Coal resources in the world, as well in Poland, are big so we can expect that it will be used for a long time, especially in the electricity sector. To do it technologically reasonably, it should be combusted in the so-called clean coal technologies. An example of one of them is the dual-fuel gas-steam technology, which couples the *Joule* gas turbine cycle with the *Clausius-Rankine* cycle of the steam turbine [1–3].

Among the possible standard gas-steam dual-fuel and coal-gas systems, there are two basic configurations [1–3, 4–6]:

- series-coupled systems, so-called Hot Windbox systems—Fig. 1.1a; the coupling takes place by means of exhaust gases from a gas turbine (these flue gases are directed as an oxidizer to the combustion chamber of a coal-fired boiler, and what is important, in this system there is no recovery boiler (heat recovery steam generator—HRSG), which is the basic device of systems coupled in parallel
- parallel-coupled systems—Figs. 1.1b, 1.4 and 1.5; coupling takes place through a steam-water system in a recovery boiler (HRSG), in which a low-temperature enthalpy of exhaust gases from a gas turbine is used (for example, the coupling consists in the production of a steam in boiler which supply the steam turbine and/or overheating, e.g. an intercooler steam in the recovery boiler, and/or heating the feed water in the exhaust-water heat exchangers built in the recovery boiler, thus eliminating steam circuit partially or completely regenerative heat exchangers; in the case of modernization of existing coal systems for dual-fuel systems, the recovery boiler powered by exhaust gases from a gas turbine is then replaced at the power plant or heat and power plant (CHP)—at least partly—requiring renovation and modernization of existing coal-fired boilers).

A. Hnydiuk-Stefan, *Dual-Fuel Gas-Steam Power Block Analysis*, Power Systems,
https://doi.org/10.1007/978-3-030-03050-6_1

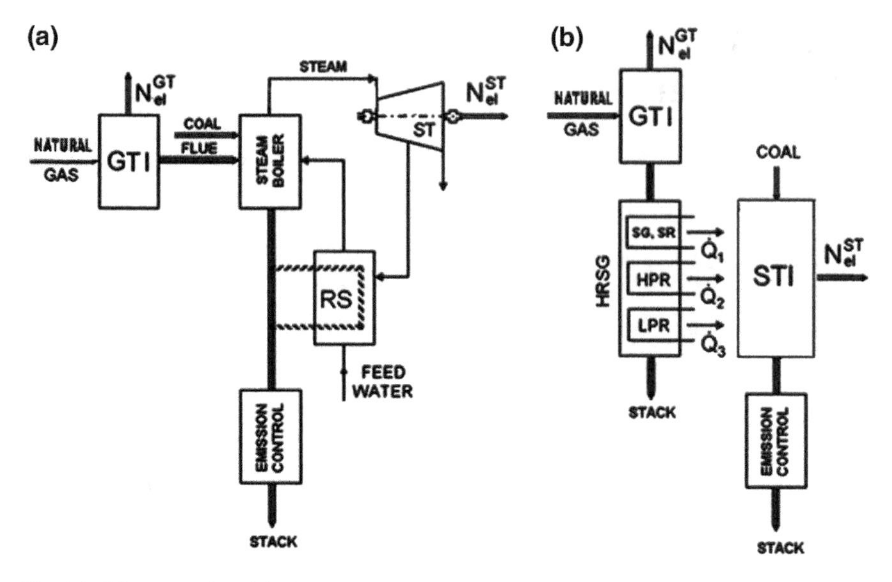

Fig. 1.1 Diagrams of dual-fuel gas-steam combined cycle: **a** series-coupled systems (Hot Windbox), **b** parallel-coupled systems. GTI—gas turbine installation, HRSG—heat recovery steam generator (recovery boiler), ST—steam turbine, STI—steam turbine installation, RS—regeneration system, SG—steam generation, SR—steam reheating, LPR—low pressure regeneration, HPR—high pressure regeneration, N_{el}^{GT}—power of gas turbine, N_{el}^{ST}—power of steam turbine

It should be mentioned that dual-fuel gas-steam systems are not so-called hierarchical systems, which are single-fuel gas-steam systems. In dual fuel systems, fuel is supplied to each of the circuits, whereas in the hierarchical system fuel is supplied only to the circulation operating in the range of the highest temperatures—Figs. 1.2 and 1.3.

For each of the subsequent cycles in a hierarchical system, operating in ever lower temperature ranges, the driving heat brought to them is the heat derived from the circuits in the hierarchy just above them. So far, only two-cycles hierarchical systems have been technically possible and are being built—Fig. 1.3 [2].

Hierarchical single-fuel gas and steam systems are currently characterized by the highest efficiency of electricity generation, for which the net efficiency exceeds even the value of 60%. Such high efficiency is the result of a very significant, about threefold increase in the temperature range of the block by combining two cycles in the hierarchical system: *Joule's* high-temperature gas turbine with the *Clausius-Rankine* low-temperature steam turbine. It is worth noting that steam pressure in hierarchical gas-steam cycles is only about 10 MPa, and not as, for example, 28 MPa in blocks for supercritical steam parameters [2], in which only the *Clausius-Rankine* cycle is implemented. The efficiency for *Clausius-Rankine's* supercritical cycles is around 50%. As for the efficiency of dual-fuel gas-steam systems, it can exceed even 50% for the operation of the steam section with subcritical parameters of fresh steam. The amount of this efficiency depends on the gas turbine used in the power

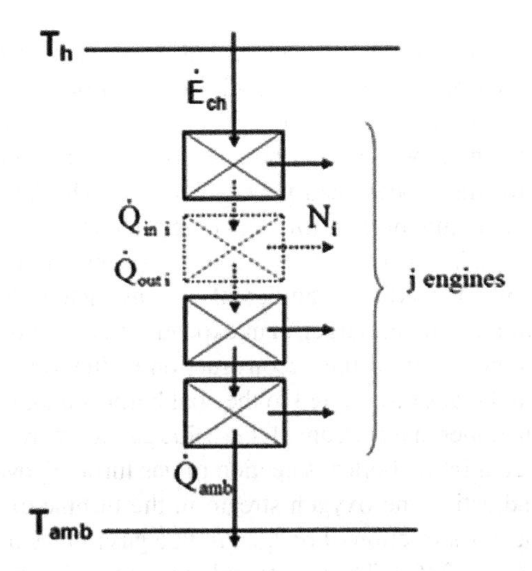

Fig. 1.2 Diagram of hierarchical j-cycle heat engine; j—number of circulating media (engines), $\dot{E}_{ch}$—stream of chemical energy of the fuel, N_i—capacity of an actual engines,$\dot{Q}_{in\ i}$, $\dot{Q}_{out\ i}$—heat of stream input into and output from an i-th ($i = 1 \div j$) cycle (engine), $\dot{Q}_{amb}$—stream of heat transmitted from the system into the environment, T_{amb}—absolute ambient temperature, T_h—absolute temperature combustion of the fuel

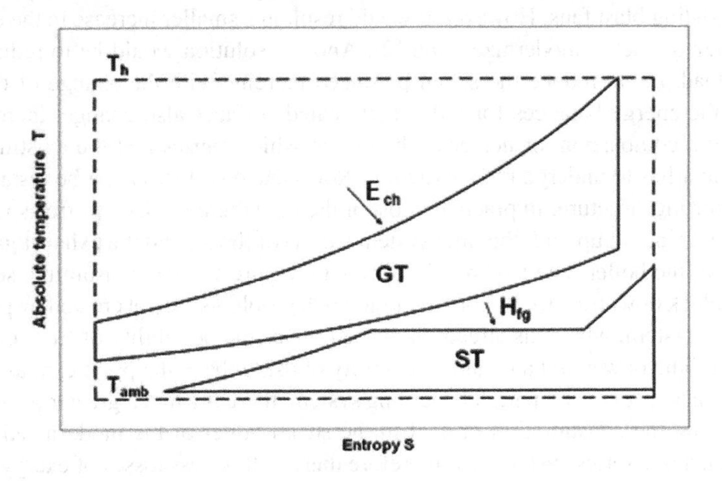

Fig. 1.3 Hierarchical cycle (theoretical) of a single-fuel gas–steam system (GT—Joule's cycle of the gas turbine, ST—Clausius-Rankine cycle of the steam turbine, E_{ch}—chemical energy of fuel transferred into GT, H_{fg}—enthalpy of flue gas exiting from the gas turbine transferred to ST through heat recovery steam generator; dashed line marks Carnot cycle for the extreme temperatures T_{amb} and T_h)

system (Sect. 4.1.1). Dual-fuel gas-steam systems, especially the parallel system, next to the construction of new power units also constitute a large modernization potential for already existing combined heat and power plants and coal-fired power plants [2, 4]. Bi-fueling systems can be created as a result of the superstructure of already existing thermal coal structures with a gas turbine. It should be noticed that in comparison with the parallel-coupled system, which is characterized by the heat recovery steam generator (recovery boiler) a big latitude in choosing the power of the gas turbine for the modernized power block, the Hot Windbox system has significantly lower economic profitability and bigger technical limitations. They are a consequence of the lack of a recovery boiler in the system, as a result of which the existing coal-fired boiler needs to be significantly reconstructed. Capital expenditures for this reconstruction are much larger than expenditures for a recovery boiler in a parallel system. The necessity of this reconstruction is due to the high temperature exhaust outlets from the gas turbine fed to the coal burners and, what is particularly important, the much higher mass stream of these flue gas compared to the replaced air stream for burning coal in the boiler (selection of gas turbine power for a particular boiler consists on adjusting the oxygen stream in the turbine exhaust gases to the boiler's demand, the mass fraction of oxygen in flue gas ranges from approx. $g_{O_2} =$ 13–16%, in the air $g_{O_2} = 23\%$). The associated increase in the speed of fumes, even with a reduced coal consumption at the same time, creates a great erosive hazard for heated surfaces. Therefore, in the Hot Windbox system, a gas turbine with relatively low power should be selected (in practice the ratio N_{el}^{GT}/N_{el}^{ST} would have to be less than 0.6) and oxygen deficiency would then be supplemented with atmospheric air using existing blast fans. However, it would result in a smaller increase in the energy efficiency of such a modernized unit [2]. Another solution would be to reduce the boiler load, i.e. to reduce the steam produced therein. With the change of the gas speed, the energy balances for individual heated surfaces also change, there is no need for a combustion air heater in the boiler, which causes that the existing coal fired boiler has to undergo reconstruction. New heated surfaces must be installed in its supporting structure. In practice, most of the time there is a lack of free space for the gas turbine set up with the inlet system (for it) of the air and the exhaust gas (out of it) near the boiler. The Hot Windbox system requires a time-consuming, several-month block downtime for its construction. Such problems are not created by parallel coupling system, which, as already indicated, gives the possibility of free selection of gas turbine power and a greater possibility of using the enthalpy of exhaust gases from it. In addition, with parallel coupling system, there is a much greater possibility of reducing the consumption of coal in the steam boiler of the modernized block rather than in a series system, and therefore there will be less losses of exergy in the system, which is very important. Therefore, the efficiency of the modernized power unit will be much higher than in the modernized serial system [1, 2]. In addition, the need to reconstruct the steam-water system in the existing coal part, and thus the financial resources for this purpose are also much smaller. The necessary investment outlays for modernization will be spent only on the newly emerging gas system and connection with the existing system. The construction of the gas system will take place at the time when the coal system "operates". Therefore, there will be no economic losses associated with its stoppage. In addition, the connection of the gas part with the carbon part can last only a few dozen days. Thus, the parallel system is in practice an energy-efficient and economically more effective way of

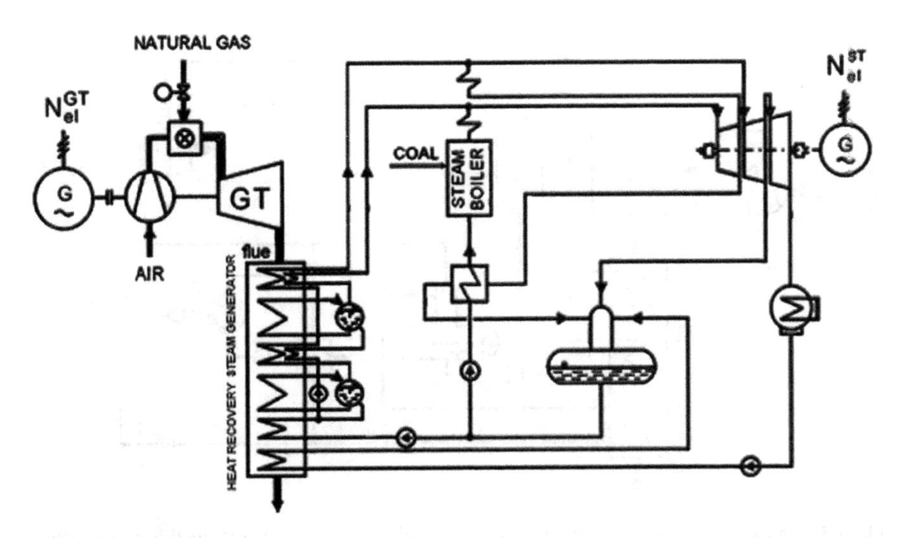

Fig. 1.4 An exemplary schematic diagram of a dual-fuel gas-steam power plant in a parallel system

modernizing coal-fired power units. What's more, at the same time, and what should be strongly emphasized, there will be even more than doubling the electric power of such modernized units, which will exclude the need to build completely new generation capacities. Therefore, what is very important, there will be a lack of the large social, economic, ecological, technological and technical problems associated with their location and construction. The emission of pollutants into the natural environment per unit of electricity generated therein will also be significantly reduced as a result of reduced coal consumption and natural gas combustion. The emission of carbon dioxide per MWh of electricity produced will decrease by as much as half to the value of around $EF_{CO_2} = 500\ kg_{CO_2}/MWh$ (Sect. 4.1.1). In Figs. 1.4 and 1.5 examples of thermal diagrams of dual-fuel gas-steam units in a parallel system are shown. Of course, there are many other technical solutions for the steam-water system in the recovery boilers coupling the gas part of the blocks (gas turbine sets) with steam circuits (coal-fired boilers and steam turbines). A general diagram of parallel circuits showing all possibilities of coupling *Joule* and *Clausius-Rankine* circuits is shown in Fig. 1.1b.

When the combustion of coal in dual-fuel gas-steam systems will take place with the simultaneous production of electricity and useful heat, for example for municipal heating needs, this will additionally lead to a significant reduction in global coal consumption and thus a reduction of harmful combustion products to the environment. Therefore, it is necessary to develop and present mathematical models and carry out technical and economic analyzes not only for dual-fuel gas-steam power plants, but also for combined heat and power plants, in order to determine the necessary conditions for the economic profitability of their application. The monograph analyzes the newly built combined heat and power plant. At the same time, it should be noted that the parallel system is also a rational way of modernization

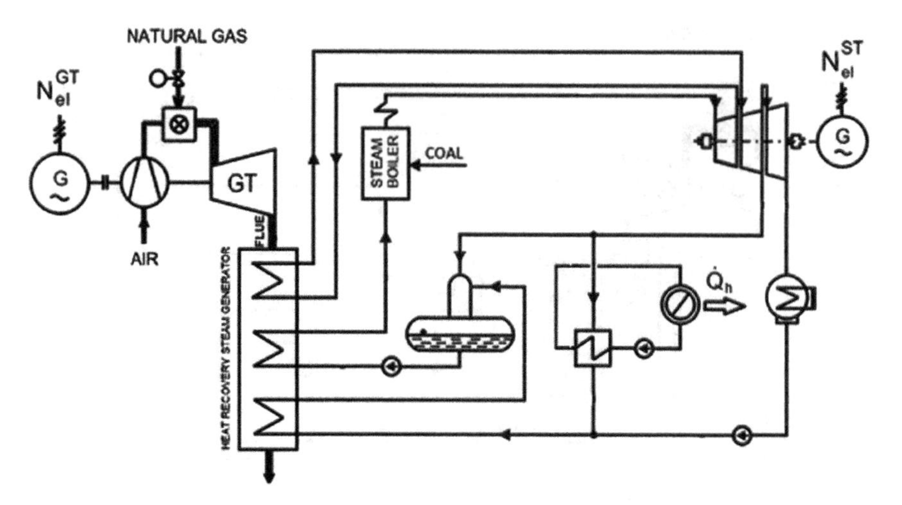

Fig. 1.5 An example of a schematic diagram of a dual-fuel gas-steam combined heat and power plant in a parallel system with an extraction-condensing steam turbine

of existing power plants and coal-fired power plants for dual-fuel systems, whereas the serial system (Hot Windbox) is economically justifiable only for newly built power plants and combined heat and power plants. The modernization of the existing CHP plant and power plant for the serial system (it should be remembered that there is no recovery boiler in the series system) would require a completely new, very expensive investment coal-fired boiler for the combined heat and power plant and power plant, whereas modernization to the parallel system requires cheap recovery boiler [1–3]. The coal boiler is the most expensive device in power units, both single and dual fuel. It is many times more expensive than a gas turbine set, while in the meantime the recovery boiler is about 2.5 times cheaper than a gas turbine set.

In order for the use of gas-steam systems to be economically viable, the price of electricity must be high enough in relation to the price of natural gas burnt in a gas turbine [2]. The current relation of this price to the price of expensive gas is too low and gas-steam systems are usually economically unprofitable. This profitability can be higher by increasing this relationship, but not only. For bi-fuel systems, it can also be increased by reducing CO_2 emissions from them, which is possible due to the greater combustion of gas in it, increasing the power of the gas turbine and thus reducing the consumption of coal. CO_2 emission from natural gas combustion is about 2 times lower per unit of chemical energy of the fuel burned than emissions from coal combustion (for hard coal, CO_2 emissions are equal to $\rho_{CO_2}^{coal} \approx 95\ \mathrm{kg_{CO_2}/GJ}$, for natural gas $\rho_{CO_2}^{gas} \approx 55\ \mathrm{kg_{CO_2}/GJ}$).

The cost-effectiveness of using gas-steam, hierarchical and dual-fuel systems also artificially increases the purchase price of CO_2 emission allowances, which strongly push the EU's "fifteen" countries. However, it should be strongly emphasized that this increase will financially destroy energy sector and industry based on coal in Poland.

Coal is the natural wealth of Poland and guarantees the security and independence of Poland (for example, the wealth of Norway is excellent hydrogeological conditions and therefore its energy is 99% hydro).

Increasing the purchase price of CO_2 emission allowances, however, reduces the profitability of power plants and combined heat and power plants, in which only relatively cheap coal is burned (in Poland, about 3 times cheaper per unit of energy from natural gas). The cost of purchasing carbon dioxide emission allowances for the combustion (co-firing) of gas in a power plant and combined heat and power plant with the same amount of electricity and heat produced will therefore be lower. How much? It depends on the purchase price of these allowances. There is therefore a limit value of this price, above which, given the prices of electricity, heat and fuels, a reduction in CO2 emissions will make the use of gas and steam systems profitable.

Mathematical models developed and presented in the monograph, which are essential, allow continuous analysis of the price relationships between fuel prices, electricity, heat prices and the price of CO_2 emission allowances, for which it is profitable to use dual-fuel gas-steam systems. Moreover, and what is very important because it is the foundation and the essence of these systems, enable the selection of the optimal power of the gas turbine set, i.e. the power that guarantees the lowest costs of electricity and heat production.

References

1. Bartnik R (2013) The modernization potential of gas turbines in the coal-fired power industry. Thermal and economic effectiveness. Springer, London
2. Bartnik R (2009) Gas-steam power plants and combined heat and power plants. Energy and economic efficiency (in Polish: Elektrownie i elektrociepłownie gazowo-parowe. Efektywność energetyczna i ekonomiczna). WNT, Warszawa (reprint 2012, 2017)
3. Bartnik R, Bartnik B (2014) Economic calculation in the energy sector (in Polish: Rachunek ekonomiczny w energetyce). WNT, Warszawa
4. Bartnik R, Buryn Z, Hnydiuk-Stefan A, (2018) Methodology and mathematical model with the continuous time for the selection of the optimal power of the gas turbine set for the dual fuel gas-steam combined cycle in a parallel system. Appl Therm Eng 141:1161–1172. https://doi.org/10.1016/j.applthermaleng.2018.06.046
5. Bartnik R, Skomudek W, Buryn Z, Hnydiuk-Stefan A, Otawa A, (2018) Methodology and Continuous Time Mathematical Model to Select Optimum Power of Gas Turbine Set for Dual-Fuel Gas-Steam Combined Heat and Power Plant in Parallel System. Energies 11(7): 1784. https://doi.org/10.3390/en11071784
6. Hnydiuk-Stefan A, Składzień J, (2015) The analysis of parameters of the cryogenic oxygen unit cooperating with power plant to realize oxy-fuel combustion. Arch Thermodyn 36(1): 39–54. https://doi.org/10.1515/aoter-2015-0003

Chapter 2
Methodology with Continuous Time Analysis of the Technical and Economic Viability of Dual-Fuel Gas and Steam Systems

Engaging financial resources in any business venture should be preceded by finding the optimal investment strategy. For example, investments in the energy sector require finding answers to the following questions.

- What technologies should be used?
- What impact do the prices of energy carriers and the relations between them have on the final value of the assumed goal criterion when looking for the optimal investment strategy?
- How to spread out own or credited financial funds during the time horizon to achieve the planned goal?

The above questions are questions about the economic efficiency of investments in the energy sector. It is obvious that it should be as large as possible that the costs of generating electricity and heat should be as low as possible.

In order to be able to examine any technical, economic, technical-economic phenomenon, etc., it is necessary to build its mathematical model, ie a mathematical record describing its functional space. For it is a great timeless truth that: *It is impossible to know the things of this world if you do not know their mathematics*, which Roger Bacon (1214–1294) already spoke in the 13th century.

Mathematical models are investigated by functional analysis, i.e. a mathematical analysis department dealing with the study of functional space properties. The word *functional* means a function whose argument is also a function, not a number. A particularly important department of functional analysis is the variational calculus, which deals with the search for extremes of functionals, which in the study of mathematical models is important.

A special case of finding extremes of functionals (the so-called *Mayer* problem) is the *Lagrange issue*, ie the search for extremes (maxima and minima) of integral functions

A. Hnydiuk-Stefan, *Dual-Fuel Gas-Steam Power Block Analysis*, Power Systems,
https://doi.org/10.1007/978-3-030-03050-6_2

$$J = \int_{t_O}^{t_F} F[x_1(t), x_2(t), \ldots, x_n(t); u_1(t), u_2(t), \ldots, u_n(t); t]\, dt = \text{extreme} \quad (2.1)$$

where:

$x_i\ (t)$ dependent variables, so-called state variables ($i = 1, 2, \ldots, n$); state variables are coordinates the state vector $\boldsymbol{x}(t) = [x_1(t), x_2(t), \ldots, x_n(t)]$,

t independent variable (for example time), $t \in \langle t_O, t_F \rangle$,

wherein:

$$\frac{dx_i}{dt} \equiv u_i(t). \quad (2.2)$$

The *Lagrange* issue is also a special case of the so-called the *issues of optimal control*, which consists in determining r control variables $u_k = u_k(t)$ $(k = 1, 2, \ldots, r)$ extremalizing integral function (goal criterion)

$$J = \int_{t_O}^{t_F} F[x_1(t), x_2(t), \ldots, x_n(t); u_1(t), u_2(t), \ldots, u_r(t); t] \quad dt = \text{extreme} \quad (2.3)$$

where the derivatives of the function x_i(t) satisfy in this case n first order differential equations called state equations

$$\frac{dx_i}{dt} = f_i[x_1(t), x_2(t), \ldots, x_n(t); u_1(t), u_2(t), \ldots, u_r(t); t]. \quad (2.4)$$

and control $u_k(t)$ are coordinates of the control vector $\boldsymbol{u}(t) = [u_1(t), u_2(t), \ldots, u_r(t)]$.

Differential Eq. (2.4) describe the changes occurring in time presented by the function (2.3) of the process.

Optimal strategies, i.e. the extremising functional (2.3) functions of controls $u_k(t)$, determine the optimal trajectory $x_i = x_i(t)$ in the n-dimensional state space.

In practice, there is a large class of technical and economic tasks, where instead of a continuous-time function (2.3), whose evolution is described by differential Eq. (2.4), we have to do with processes that are essentially discrete in their own right. These include mainly multi-step tasks to make decisions. This is the economic process, which is usually described by differential equations. The step of discretization is then determined by its cycle. In practice, this is usually a year, $\Delta t = 1$ year. In this case, when looking for the extreme of the target function (goal criterion) with the number of considered process steps equal to N

$$J = \sum_{t=1}^{N} F[x_1(t), x_2(t), \ldots, x_n(t); u_1(t), u_2(t), \ldots, u_r(t); t]$$
$$= \text{extreme}, (N = 1, 2, \ldots) \quad (2.5)$$

with differential equations of states

$$x_i(t-1) = f_i[x_1(t), x_2(t), \ldots, x_n(t); u_1(t), u_2(t), \ldots, u_r(t); t] \quad (i = 1, 2, \ldots, n) \tag{2.6}$$

Bellman's principle of optimality can be used. This rule expresses the fact that each part of the optimal trajectory optimizes the function for the respective starting and ending points. In other words, for the trajectory to be optimal, each part (step) must be optimal, regardless of the starting point from which it emerges. This means that the search for the optimal control should be carried out for each step $t = 1, \ldots, N$ separately, with the corresponding initial state, i.e. the extreme value resulting from the previous step, with the extremes being determined in each step according to the bonds occurring in it, using the *Lagrange* multipliers method for this purpose. *Bellman's* approach, therefore, allows to search for the extremum of the function by searching for extremes of function and leads to a recursive formula expressing N *Bellman equations*, in which the letter S denotes extreme functions for steps $t - 1$ and t

$$\begin{aligned} S[x_1(t), x_2(t), \ldots, x_n(t); t] = {} & F[x_1(t), x_2(t), \ldots, x_n(t); u_1(t), u_2(t), \ldots, u_r(t); t] \\ & + S[x_1(t-1), x_2(t-1), \ldots, x_n(t-1); t-1], \\ & t = 1, \ldots, N \end{aligned} \tag{2.7}$$

whereby, to obtain a uniform record of the formula (2.7), it was assumed that $S[x_1(0), x_2(0), \ldots, x_n(0); 0] = 0$. The S functions in formula (2.7) are obtained with the use of optimal controls $u_k = u_k(t)$ $(k = 1, 2, \ldots, r)$ determined by using the r system of functional Eq. (2.10).

Determining the functional equation (2.7) in relation to the function $S[x_1(t), x_2(t), \ldots, x_n(t); t]$ is tantamount to stepwise constructing a class of optimal strategies for many initial states. This task, with the number of dependent variables $x_i(t)$, even just above two is very extensive. This abundance causes that approximate methods are often used to solve specific problems. Also for issues that are less "extensive" incomparably faster can be obtained by the approximate method even in situations where high accuracy of calculations is required. An effective method is to replace then the differential equations of state (2.6) and *Bellman's methodology* (2.7) with continuous function (2.3) and differential equation (2.4), and then solve them for example with the approximate *Ritz* method known from the variational calculus [1, 2].

Equations (2.6) and (2.7) represent ordinary (front) recursion. When in the search for the optimal trajectory, "aiming" at the final point N with the values of $x_i = x_i(N)$, then Eqs. (2.6) and (2.7) should be replaced by retrograde recursion

$$x_i(t+1) = f_i[x_1(t), x_2(t), \ldots, x_n(t); u_1(t), u_2(t), \ldots, u_r(t); t] \tag{2.8}$$

$$S[x_1(t), x_2(t), \ldots, x_n(t); t] = F[x_1(t), x_2(t), \ldots, x_n(t); u_1(t), u_2(t), \ldots, u_r(t); t] + S[x_1(t+1), x_2(t+1), \ldots, x_n(t+1); t+1], \quad t = N-1, \ldots, 0 \tag{2.9}$$

The retrograde recursion, which in its essence is characterized by the retrograde logic of inference, is, therefore, what is extremely important, a way of analyzing the future. It creates a scientific way of thinking about it. It begins with the assumption of the desired value in the final year N and then goes back step by step, to calculate the value that should be in the present moment, so that it could be achieved in the year N this assumed desirable value. This undoing and calculation is also carried out after the optimal trajectory for the assumed goal criterion, and with different assumed control scenarios. Scenarios are conditioned, for example, by changes in the prices of energy carriers, unit rates of charges for emission of pollutants into the environment, etc. The calculated present value shows what technology and its technical solutions should be adopted today so that the intended goal can be achieved in the future after an optimal trajectory. Therefore, it enables analysis and selection of alternative solutions together with all their associated conditions. In other words, it allows for a broad analysis of various investment scenarios and technologies used in various scenarios of the evolution of energy carrier prices and environmental conditions, in order to achieve the desired final value in the assumed time horizon of N years.

As mentioned above, finding the trajectory of the optimal goal criterion (2.5) comes down to solving (for each of the next steps t) the system of r function equations to determine r controls $u_k = u_k(t)$ (k = 1, 2, ..., r) extremating the function (2.5) in a given step t. This system is as follows:

$$\begin{cases} \dfrac{\partial\{F[x_1(t), x_2(t), \ldots, x_n(t); u_1(t), u_2(t), \ldots, u_r(t); t] + S[x_1(t \mp 1), x_2(t \mp 1), \ldots, x_n(t \mp 1); t \mp 1]\}}{\partial u_1} = 0 \\ \dfrac{\partial\{F[x_1(t), x_2(t), \ldots, x_n(t); u_1(t), u_2(t), \ldots, u_r(t); t] + S[x_1(t \mp 1), x_2(t \mp 1), \ldots, x_n(t \mp 1); t \mp 1]\}}{\partial u_2} = 0 \\ \cdots\cdots\cdots\cdots\cdots\cdots\cdots\cdots\cdots\cdots \\ \dfrac{\partial\{F[x_1(t), x_2(t), \ldots, x_n(t); u_1(t), u_2(t), \ldots, u_r(t); t] + S[x_1(t \mp 1), x_2(t \mp 1), \ldots, x_n(t \mp 1); t \mp 1]\}}{\partial u_r} = 0 \end{cases} \tag{2.10}$$

in which the extreme function S was determined in the previous step $t \mp 1$ (in the case of forward recourse in step $t - 1$, in the case of reverse recursion in step $t + 1$). In the system (2.10), before the partial derivations $\partial(F + S)/\partial u_k$ are determined as variables $x_i(t \mp 1)$ in the S function, the differential equation (2.6) should be substituted in the case of the forward recursion, in the case of the reverse recursion of the Eq. (2.8).

2.1 A Target Function with a Continuous Time When Searching for the Optimal Investment Strategy in the Energy Sector

Until now, the measures for assessing the economic effectiveness of any investment projects were presented in the literature of the subject only by means of discrete records, using geometrical series (the discount calculation is essentially geometrical progress), and they are used only in this form. Thus, the total discounted net profit achieved by all years of company operation is defined by the formula [1, 2]:

$$NPV = \sum_{t=1}^{N} \frac{CF_{A,t,net}}{(1+r)^t} - J_0, \tag{2.11}$$

and using it assuming that $NPV = 0$, *IRR* and *DPBP* measures are defined [1, 2]:

- internal rate of return (interest rate that the invested capital J brings; *Internal Rate of Return*)

$$\sum_{t=1}^{N} \frac{CF_{A,t,gross}}{(1+IRR)^t} = J_0, \tag{2.12}$$

- dynamic payback time of invested capital J (*Discounted Pay Back Period DPBP*)

$$\sum_{t=1}^{DPBP} \frac{CF_{A,t,net}}{(1+r)^t} = J_0, \tag{2.13}$$

wherein:

$CF_{A,t,net}$ annual net cash flows in the subsequent years $t = 1, 2, \ldots, N$ (Fig. 2.1), being the difference between the annual revenues of S_A from sales of products (e.g. electricity) and expenses (operating costs K_e, formula (2.34), and income tax P [formula (2.15)] from the annual gross profit Z_A [formula (2.15)], K_e costs, of course, do not include depreciation costs, they are not expenditure during the exploitation period; depreciation in formulas (2.11)–(2.13) is of course J_0);

$$CF_{A,t,net} = S_{A,t} - K_{e,t} - P_t, \tag{2.14}$$

where the income tax P_t is expressed in the form [see formula (2.25)]:

$$P_t = Z_A p = (S_{A,t} - K_{e,t} - \rho J_0)p \tag{2.15}$$

$CF_{A,t,gross}$ gross cash flow; gross cash flows do not include income tax P

$$CF_{A,t,gross} = S_{A,t} - K_{e,t}, \tag{2.16}$$

where:

J_0 discounted for the moment of commencing the operation of the enterprise t = 0 (Fig. 2.1) capital expenditure J incurred for its construction [1, 2]

$$J_0 = zJ \tag{2.17}$$

(expenditures $J_0 = zJ$ must, of course, be reimbursed, i.e. amortized),

N the calculation period of the company's lifetime expressed in years (Fig. 2.1),
r discount rate (the interest rate on investment capital allows for taking into account the change in the value of money over time),
$(1 + r)^{-t}$ money discounting factor,
t subsequent years of operating the enterprise, $t = 1, 2, \ldots, N$ (Fig. 2.1),
p income tax rate on annual gross profit Z_A,
ρ interest-bearing depreciation rate [1, 2]

$$\rho = \frac{r(1 + r)^N}{(1 + r)^N - 1} \tag{2.18}$$

z discounting factor (freezing ratio) investment capital J on the moment of construction completion (Fig. 2.1), $z > 1$; this coefficient is included undesirable impact of freezing investment outlays during construction, they do not bring profits in this time, but interest on capital J increases [1, 2].

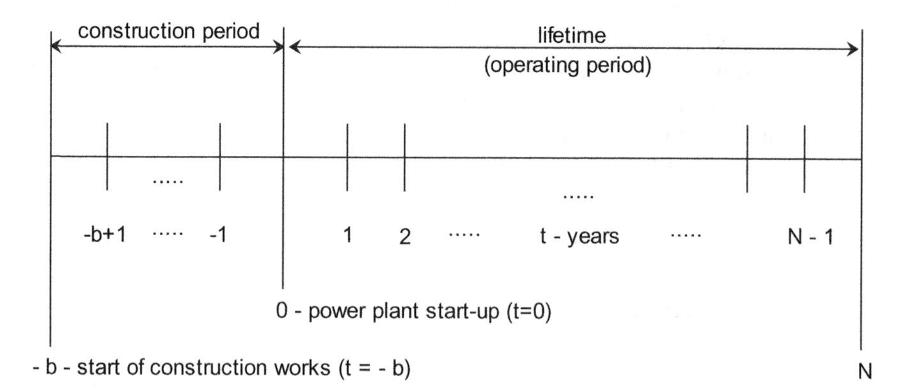

Fig. 2.1 Time diagram of the implementation of the investment project

$$z = \frac{(1+r)^{b+1} - 1}{(b+1)r}, \tag{2.19}$$

b time duration of construction expressed in years (Fig. 2.1).

In searching for the optimal investment strategy, the *NPV* net value, which is defined by the formula (2.11), should be assumed as the criterion of the target (the criterion of optimality). This formula, as shown in [1], can also be presented in the form of:

$$NPV = \sum_{t=1}^{N} \frac{S_{A,t} - K_{e,t} - F_t - R_t - \left(S_{A,t} - K_{e,t} - F_t - A_t\right)p}{(1+r)^t}, \tag{2.20}$$

and hence in the continuous time recording the total discounted *NPV* net profit is presented as functional:

$$NPV = \int_0^T \left[S_A(t) - K_e(t) - F(t) - R(t) - (S_A(t) - K_e(t) - F(t) - A(t))p\right] e^{-rt} dt \tag{2.21}$$

where:

A depreciation installment,
e^{-rt} money discounting factor,
F time-varying interest rates (financial costs) on investment funds J_0; interest F is an unknown function of the variable in installments R; $F = F[R(t)]$,
K_e time-varying annual operating costs,
p rate of income tax variable in time,
R loan repayment installment variable over time,
r discount rate variable over time,
S_A annual revenue variable in time [formulas (2.31), (2.33)],
t time,
T calculation period of the plant's operation expressed in years [the time interval T is identical to the years N from the discrete record (2.11), (2.20)]

From the formula (2.21) from the condition $NPV = 0$, further measures of the economic efficiency of the investment are determined in the continuous time record: *IRR* interest rate, what the invested capital J brings and expressed in years the time of its return *DPBP* [see formulas (2.12), (2.13)]

$$\int_0^T (S_A - K_e) e^{-IRRt} dt = \int_0^T [F(IRR) + R(IRR)] e^{-IRRt} dt, \tag{2.22}$$

$$\int_0^{DPBP} \left[S_A - K_e - (S_A - K_e - F - A)p\right] e^{-rt} dt = \int_0^T (F + R) e^{-rt} dt. \tag{2.23}$$

The *IRR* measure according to its definition [1, 2] is determined assuming that income tax P on gross profit Z [compare with the formula (2.15)]

$$Z = S_A - K_e - F - A \tag{2.24}$$

is equal to zero [see formula (2.15)]:

$$P = (S_A - K_e - F - A)p = 0. \tag{2.25}$$

The *F(IRR)* and *R(IRR)* records on the right side of the formula (2.22) mean that the financial cost F and the installment of the R loan are *IRR* functions, while on the left-hand side these values, as well as in the formulas (2.21) and (2.23) together with depreciation A are the functions of the rate r. The right sides of the formulas (2.22) and (2.23) represent the discounted capital expenditure J_0 [see formulas (2.11)–(2.13)].

In formulas (2.21)–(2.25), all functions of changes in their values may be assumed for all sub-values, e.g. any scenarios of changes in the prices of energy carriers and unit rates for emissions of pollutants into the natural environment. The record of the continuous time criterion of optimality (2.21) has, therefore, a significant advantage over discrete recordings (2.11) and (2.20). It allows to quickly and easily analyze changes in the value of *NPV* profit in order to find its largest value including any temporary changes in the value of which it is a function. What's more, it allows to study the variability of *NPV* functions and to draw up a graph using the differential calculus, which allows to obtain a whole range of additional, important information that could not be seen without, or at least would be difficult to see. It makes it possible to explicitly assess the impact of particular input quantities on final results, and above all to easily and quickly find not only the optimal solution, but also the area of solutions close to the optimal one. What's more, it allows to show the nature of their changes. It allows to discuss and analyze research results. In technology, in economics, in applications it has a large and significant value. What's more, mathematical models with continuous time allow to draw conclusions of a general nature, and only the path from the general to the detail is correct and gives the opportunity to generalize the considerations.

The transition from the detail to the general most often—not to say that usually—is not true. The methodology and obtained mathematical models thanks to the use of the *NPV* measure record by means of integral functionality (2.21) enable, which is impossible with the use of its discrete record, wide technical and economic analysis of any investment projects, including energy-related ones, with any scenarios of changes in time of all sizes sub-characteristic describing them. This methodology is, therefore an important way of analyzing the future. It creates thinking about it in a scientific way.

Summing up, the innovative methodology developed and applied in the monograph in a continuous time record [formula (2.21)] gives a completely new quality of technical and economic analyzes of all investment processes, because it allows for:

- use in *NPV* mathematical models obtained by means of continuous time recording (2.21) of any sub-functions (ie any time scenarios) characterizing the analyzed investment process and obtaining the *NPV* function, which gives a comprehensive picture of changes in the *NPV* value of this process; the discrete record gives only the detail, i.e. the numerical value of *NPV*, which of course prevents any analysis of the process
- thanks to the received *NPV* function it is possible to draw conclusions of a general nature, and only the path from general to specific is correct and gives the opportunity to generalize the considerations, while the transition from detail to the general—not to say that usually—is not true
- using the differential calculus to find the largest value of the *NPV* function
- examining the variability of the *NPV* function in order to obtain comprehensive information about it, as well as to prepare a graph, which allows obtaining a whole range of additional, important information that would be impossible without it, at least it would be difficult to see
- discussion and analysis of research results; in technology, in economics, in applications it has a large and significant value.

The choice of the optimal investment strategy should be made at:

$$NPV \rightarrow max \tag{2.26}$$

for the assumed power value of the N_{el} power plant. The amounts to be optimized (decision variables) are:

- available technologies and within a given technology
- technical solutions, and thus the devices used, their constructional and rated performance parameters, the structure of their connections, process operating parameters, etc.

In the general case, finding the extreme of integral functional (2.20) or (2.21) with their associated state equations [e.g., formula (2.29)] and equations of constraints [e.g. formula (2.27)], comes to finding sub-integral functions that would limit this function. To find them, the *Bellman principle* of optimality or *Pontryagin's maximum* principle should be used for this purpose [1, 2]. The first of these applies to discrete processes (2.20) and the other continuous time processes (2.21). In a situation when sub-functions are assumed in advance (the so-called direct method of solving variational problems, e.g. the *Ritz* method [1, 2]), of course, to meet the boundary conditions imposed on them, the task trivializes. After integrating the dependence (2.21) within given limits, finding the extreme comes down only to the determination of constants (also called controls [1, 2]) occurring in assumed sub-functions (in the problem discussed in the chapter these are constants: a_{el}, a_{fuel}, a_{CO_2}, a_{CO}, a_{SO_2}, a_{NO_X}, a_{dust}, b_{CO_2}, formulas (2.32), (2.36), (2.38)–(2.42), (2.44). For this purpose, the necessary conditions for the existence of an extreme should be used, ie the partial derivation obtained by integration of functions relative to these constants [formula (2.40)]. Obviously, these conditions are constituted by n equations, where the number n equals the number of constants.

The continuous record (2.21) of the criterion of optimality has an incomparable advantage over discrete recording (2.20). It allows to quickly and easily analyze the changes in the value of NPV profit in order to find its largest value. A single operation of integrating dependencies (2.21) gives a compact form of the NPV formula [formula (2.40)], which is convenient for such analysis, while the time-consuming and "extensive" step-by-step calculation process in subsequent years $t = 1, 2, \ldots,$ N consecutive values of the functional (2.20) and their summation, this option does not give. However, in order to integrate the function (2.21), all of the sub-functions, as already stated above, ie S_R income, operating costs K_e, financial cost F, loan installment R and depreciation amortization A must be known functions of time t. Otherwise, i.e. when integer expressions are unknown functions, finding the maximum value of functionals (2.20) or (2.21) requires the *Bellman optimality principle* or the *Pontriagin* maximum rule.

The power value of the N_{el} power plant in the optimization calculations should not change. The investment process of building a power plant consists of many tasks: from the process of obtaining a building permit, the process of obtaining the source of investment financing, the design process and finally the construction process. It is therefore a long-term process and therefore in all years (steps) $t = 1, 2, \ldots, N$ or in the whole calculation period $t \in \langle 0,\ \mathrm{T}\rangle$ power N_{el} in search of functional maxima (2.20) and (2.21) should be adopted as a constant value. Only the aforementioned annual revenues and annual operating costs of the power plant (operating costs plus capital costs), which depend, of course, on the technology used and the assumed N_{el} power, are only changed in time.

A limit is imposed on the sum of the R loan installments. It must be equal to the discounted investment expenditure of J_0. The condition of constraints in a continuous time record is expressed by the following formula:

$$\int_0^T R dt = J_0. \tag{2.27}$$

In practice, the loan repayment installment has a constant value $R = \text{const}$ and from the Eq. (2.27) it is obtained:

$$R = \frac{J_0}{T}. \tag{2.28}$$

The amortization depreciation A is expressed in the same form as the installment R. The outlays of J_0 depend on the applied energy technology and technical solutions as well as on the value of the assumed electrical power N_{el} of the power plant. They are therefore different for different technologies and technical solutions with the same power values.

In the general case, the evolution of the unknown function of the financial cost $F = F[R(t)]$ describes the equation of state:

$$\frac{dF}{dt} = -rR. \tag{2.29}$$

In practice, however, the repayment installment has a fixed value R = const and interest F is expressed by the function:

$$F(t) = r[J_0 - (t - 1)R]. \tag{2.30}$$

The other functions appearing in the functional are presented below (2.21).

- The revenue function

$$S_A(t) = E_{el,A}\mathrm{e}_{el}(t) \tag{2.31}$$

achieved from the sale of electricity, whereby the change in time unitary (per unit of energy) of e_{el} electricity prices can be presented, for example, using the exponential function (depending on the value a_{el}, the e_{el} price may increase, decrease or remain unchanged in subsequent years)

$$\mathrm{e}_{el}(t) = \mathrm{e}_{el}^{t=0}\mathrm{e}^{a_{el}t}. \tag{2.32}$$

In the case of combined heat and power plants, revenue $S_A(t)$ from heat sales and revenues from the sale of property rights should be taken into account in addition to revenue from electricity sales $S_{A,el} = E_{el,A}\mathrm{e}_{el}^{t=0}\mathrm{e}^{a_{el}t}$:

$$S_A(t) = E_{el,A}\mathrm{e}_{el}^{t=0}\mathrm{e}^{a_{el}t} + Q_A\mathrm{e}_{c}^{t=0}\mathrm{e}^{a_{c}t} + E_{CHP,A}\mathrm{e}_{CHP}^{t=0}\mathrm{e}^{a_{CHP}t} \tag{2.33}$$

where $E_{CHP,A}\mathrm{e}_{CHP}^{t=0}\mathrm{e}^{a_{CHP}t}$ means income from the sale of property rights resulting from the so-called certificates of origin of electricity generated in high-efficiency cogeneration

where:

a_{CHP} control (exponent $\mathrm{e}_{CHP}^{t=0}\mathrm{e}^{a_{CHP}t}$) change in price of the certificate of origin of electricity produced in high-efficiency cogeneration,

$\mathrm{e}_{CHP}^{t=0}$ the initial unit price of the certificate of origin of electricity produced in high-efficiency cogeneration,

$E_{CHP,A}$ annual amount of electricity produced in high-efficiency cogeneration in accordance with Directive 2004/8/EC of the European Parliament,

Q_A annual net production of heat, GJ/a.

In the calculations presented in the monograph, a policy dependent component $E_{CHP,A}\mathrm{e}_{CHP}^{t=0}\mathrm{e}^{a_{CHP}t}$ was omitted in formula (2.33). It was introduced by policy as support for generating electricity in cogeneration sources. Additional income from certificates is to theoretically lower the unit heat cost $k_{h,av}$. In practice, however, it should be expected that it will be the other way round, and that is why in the economic analyzes the effect of the subsidy was deliberately omitted. All artificial, economic-dependent elements of financial support of associated systems, which costs are borne by taxpayers, lead only to various kinds of perceptions, whose financial costs will eventually affect the recipient of heat. Thus, subsidies will spiral the price increase and will raise the price of heat above that which would have been without them. In addition it's one time mechanism, and in a moment it may not be available for producers. The elements of support only mix up the image of thermodynamic processes and energy technologies in society, although they are intended to be rationalized.

The variable annual operating costs of K_e include: the cost of fuel K_{fuel}, the cost of K_{sw} supplementary water, the cost of wages with cash surcharges K_{sal}, the cost of maintenance and renovation K_{serv}, the cost of non-energy raw materials and auxiliary materials K_m, the cost for economic use of the environment K_{env} (i.a. fees for emission of exhaust gases into the atmosphere, sewage disposal, storage of waste, etc.), the cost of taxes, fees and insurance K_P and the cost of purchasing permits for emission of carbon dioxide K_{CO_2}

$$K_e = K_{fuel} + K_{sw} + K_{sal} + K_{serv} + K_m + K_{env} + K_P + K_{CO_2}. \tag{2.34}$$

The cost, which is the effect of the EU climate policy, leads to a multiplication of the K_e operation costs of power plant. The sums of costs $K_{sw} + K_m$ and $K_{sal} + K_P$ in the formula (2.34) can be taken into account by increasing, for example, a few percent, the cost of K_{fuel} and a dozen or so dozens percent, the cost of K_{serv}.

- Fuel cost function

$$K_{fuel}(t) = E_{ch,A} e_{fuel}(t) \tag{2.35}$$

wherein the change in the unit price (per unit of energy) of the fuel e_{fuel} can be recorded, for example, by an equation

$$e_{fuel}(t) = e_{fuel}^{t=0} e^{a_{fuel} t}. \tag{2.36}$$

- The cost function for using the natural environment

$$K_{av}(t) = E_{ch,A}\left[\rho_{CO_2} p_{CO_2}(t) + \rho_{CO} p_{CO}(t) + \rho_{NO_x} p_{NO_x}(t) + \rho_{SO_2} p_{SO_2}(t) + \rho_{dust} p_{dust}(t)\right] \tag{2.37}$$

whereas changes in the unit rates (per unit of mass) for emissions of CO_2, CO, NO_x, SO_2 and dust can be recorded, for example, by equations

$$p_{CO_2}(t) = p_{CO_2}^{t=0} e^{a_{CO_2} t} \tag{2.38}$$

$$p_{CO}(t) = p_{CO}^{t=0} e^{a_{CO} t} \tag{2.39}$$

$$p_{NO_X}(t) = p_{NO_X}^{t=0} e^{a_{NO_X} t} \tag{2.40}$$

$$p_{SO_2}(t) = p_{SO_2}^{t=0} e^{a_{SO_2} t} \tag{2.41}$$

$$p_{dust}(t) = p_{dust}^{t=0} e^{a_{dust} t}. \tag{2.42}$$

- The function of the cost of purchasing CO_2 emission allowances

$$K_{CO_2}(t) = E_{ch,A}(1 - u)\rho_{CO_2} e_{CO_2}(t) \tag{2.43}$$

whereas the change in the unit price (per unit of mass) of CO_2 emission allowances purchase can be recorded, for example, by the equation

$$e_{CO_2}(t) = e_{CO_2}^{t=0} e^{b_{CO_2} t}, \tag{2.44}$$

where:

$a_{el}, a_{fuel}, a_{CO_2}, a_{CO}, a_{SO_2}, a_{NO_X}, a_{dust}, b_{CO_2}$	controls,
$E_{el,A}$	annual net electricity production,
$E_{ch,A}$	annual consumption of chemical fuel energy,
u	share of chemical fuel energy in its total annual consumption, for which it is not required to purchase CO_2 emission allowances,
$\rho_{CO_2}, \rho_{CO}, \rho_{NO_x}, \rho_{SO_2}, \rho_{dust}$	emissions of CO_2, CO, NO_x, SO_2, dust per unit of chemical fuel energy

- The cost of maintenance and renovation function

$$K_{serv} = \delta_{serv} J \tag{2.45}$$

where:

δ_{serv} annual fixed cost rate depending on investment expenditures (maintenance costs, equipment repairs, in practice value $\delta_{serv} \cong 3\%$)

In Eqs. (2.32), (2.36), (2.38)–(2.42) and (2.44) price evolution and unit rates for emissions of pollutants into the natural environment depending on value a_{el}, a_{fuel}, a_{CO_2}, a_{CO}, a_{SO_2}, a_{NO_X}, a_{dust}, b_{CO_2} are strongly monotonic or permanent functions in time.

The above formulas (2.31), (2.35), (2.37) and (2.43) are general formulas. They concern both the annual production of electricity in a gas turbine $E_{el,A}^{GT}$ as well as steam $E_{el,A}^{ST}$ and the annual consumption of chemical energy of gas $E_{ch,A}^{gas}$ and coal $E_{ch,A}^{coal}$. Also, the individual costs of pollutant emissions to the natural environment in Eq. (2.37) are described for gas and coal with the same Eqs. (2.38)–(2.42), except that for each fuel there are different emission values ρ_{CO_2}, ρ_{CO}, ρ_{NO_x}, ρ_{SO_2}, ρ_{dust} and different values of controls a_{fuel}, a_{CO_2}, a_{CO}, a_{SO_2}, a_{NO_X}, a_{dust}, b_{CO_2}.

References

1. Bartnik R, Bartnik B, Hnydiuk-Stefan A (2016) Optimum investment strategy in the power industry. Mathematical models. Springer, New York
2. Bartnik R, Buryn Z, Hnydiuk-Stefan A (2017) Investment strategy in heating and CHP. Mathematical models. Wydawnictwo Springer, London

Chapter 3
Mathematical Models for Time Continuous Analysis of Technical and Economic Effectiveness of Newly Built Dual Fuel Gas-Steam Turbines

After substitution to (2.21) dependences from (2.28), (2.30), (2.31)–(2.45) solution for finding maximum value of functional (2.21) comes down to, as noted above, trivial task—to performing single integration within defined boundaries and analyzing volatility of resulting *NPV* function of independent variables: a_{el}, a_{pal}, a_{CO_2}, a_{CO}, a_{SO_2}, a_{NO_X}, a_{pyl}, b_{CO_2} [1, 2].

3.1 Models for Power Plant

After integration (2.21) formula for *NPV* profit for both serial and parallel power plant layout is presented by following equation:

A. Hnydiuk-Stefan, *Dual-Fuel Gas-Steam Power Block Analysis*, Power Systems,
https://doi.org/10.1007/978-3-030-03050-6_3

$$
\begin{aligned}
NPV = & \Bigg\{ (E_{el,A}^{ST} + E_{el,A}^{GT})(1 - \varepsilon_{el}) e_{el}^{t=0} \frac{1}{a_{el} - r} [\mathrm{e}^{(a_{el}-r)T} - 1] \\
& - E_{ch,A}^{gas} \Bigg\{ (1 + x_{sw,m,was}) e_{gas}^{t=0} \frac{1}{a_{gas} - r} [\mathrm{e}^{(a_{gas}-r)T} - 1] \\
& + \rho_{CO_2}^{gas} p_{CO_2}^{t=0} \frac{1}{a_{CO_2} - r} [\mathrm{e}^{(a_{CO_2}-r)T} - 1] + \rho_{CO}^{gas} p_{CO}^{t=0} \frac{1}{a_{CO} - r} [\mathrm{e}^{(a_{CO}-r)T} - 1] \\
& + \rho_{NO_X}^{gas} p_{NO_X}^{t=0} \frac{1}{a_{NO_X} - r} [\mathrm{e}^{(a_{NO_X}-r)T} - 1] + \rho_{SO_2}^{gas} p_{SO_2}^{t=0} \frac{1}{a_{SO_2} - r} [\mathrm{e}^{(a_{SO_2}-r)T} - 1] \\
& + \rho_{dust}^{gas} p_{dust}^{t=0} \frac{1}{a_{dust} - r} [\mathrm{e}^{(a_{dust}-r)T} - 1] + (1 - u) \rho_{CO_2}^{gas} e_{CO_2}^{t=0} \frac{1}{b_{CO_2} - r} [\mathrm{e}^{(b_{CO_2}-r)T} - 1] \Bigg\} \\
& - E_{ch,A}^{coal} \Bigg\{ (1 + x_{sw,m,was}) e_{coal}^{t=0} \frac{1}{a_{coal} - r} [\mathrm{e}^{(a_{coal}-r)T} - 1] \\
& + \rho_{CO_2}^{coal} p_{CO_2}^{t=0} \frac{1}{a_{CO_2} - r} [\mathrm{e}^{(a_{CO_2}-r)T} - 1] + \rho_{CO}^{coal} p_{CO}^{t=0} \frac{1}{a_{CO} - r} [\mathrm{e}^{(a_{CO}-r)T} - 1] \\
& + \rho_{NO_X}^{coal} p_{NO_X}^{t=0} \frac{1}{a_{NO_X} - r} [\mathrm{e}^{(a_{NO_X}-r)T} - 1] + \rho_{SO_2}^{coal} p_{SO_2}^{t=0} \frac{1}{a_{SO_2} - r} [\mathrm{e}^{(a_{SO_2}-r)T} - 1] \\
& + \rho_{dust}^{coal} p_{dust}^{t=0} \frac{1}{a_{dust} - r} [\mathrm{e}^{(a_{dust}-r)T} - 1] + (1 - u) \rho_{CO_2}^{coal} e_{CO_2}^{t=0} \frac{1}{b_{CO_2} - r} [\mathrm{e}^{(b_{CO_2}-r)T} - 1] \Bigg\} \\
& - J(1 - \mathrm{e}^{-rT})(1 + x_{sal,t,ins}) \frac{\delta_{serv}}{r} - zJ \left(\frac{1 - \mathrm{e}^{-rT}}{T} + 1 \right) \Bigg\} (1 - p) \rightarrow \max
\end{aligned} \tag{3.1}
$$

where:

a_{el}, a_{gas}, a_{coal}, a_{CO_2}, a_{CO}, a_{SO_2}, a_{NO_X}, a_{dust}, b_{CO_2}—exponents expressing time evolution of the price of energy carriers and charges for the emission of harmful combustion products into the environment (e.g. $e_{el}(t) = e_{el}^{t=0} \mathrm{e}^{a_{el} t}$, etc.; depending on the a_{el} value, the e_{el} price may increase, decrease or remain unchanged in subsequent years) [1, 2],

e_{el}, e_{gas}, e_{coal}, e_{CO_2}	unit price of electricity, natural gas, coal and purchase of CO_2 emission allowances,
$E_{ch,A}^{gas}$	chemical energy of gas burnt in a gas turbine during the year
$E_{ch,A}^{coal}$	chemical energy of coal burned in the boiler during the year,
$E_{el,A}^{GT}$	gross electricity produced in a gas turbo set during the year,
$E_{el,A}^{ST}$	gross electricity produced in a steam turbo set during the year,
J	capital expenditures for the plant,
p_{CO_2}, p_{CO}, p_{NO_x}, p_{SO_2}, p_{dust}	charges for the emission of harmful combustion products into the environment,
p	the income tax rate on gross profit,
r	discount rate,
T	years of block operation,

u	share of chemical fuel energy in its total annual consumption, for which it is not required to purchase CO_2 emission permits,
$x_{sw,m,was}$	factor taking into account the costs of supplementary water, auxiliary materials, sewage disposal, storage of slag, waste (in practice, the value of $x_{sw,m,was}$ is approx. 0.25),
$x_{sal,t,ins}$	factor including costs of wages, taxes, insurance, etc. (in practice, the value of $x_{sal,t,ins}$ is approx. 0.02),
z	freezing capital investment ratio,
δ_{serv}	the rate of fixed costs depending on investment expenditures (maintenance costs, overhauls of equipment),
ε_{el}	indicator of electrical own needs of the power block,
ρ_{CO_2}, ρ_{CO}, ρ_{NO_x}, ρ_{SO_2}, ρ_{dust}	CO_2, CO, NO_x, SO_2, and dust emissions per unit of chemical fuel energy.

The question should be asked if the integral function (2.21), with many unknown functions [(2.32), (2.36), (2.38)–(2.42), (2.44)] has an extreme? The answer is obvious that it don't have. It results from the unlimited possibility of increase/decrease in prices of energy carriers and environmental fees, i.e. increase/decrease in revenues [formula (2.31)] and increase/decrease in costs [formulas (2.34)–(2.45)]. It is easy to prove by solving the issue of variational searching for the extreme of a functional (2.21) by means of the *Ritz method*. All functions of revenues and individual costs in the formula (2.21) are functions that grow in the whole range $\pm\infty$ of changes in the values of independent variables a_h, a_{el}, a_{fuel}, a_{CO_2}, a_{CO}, a_{SO_2}, a_{NO_X}, a_{dust}, b_{CO_2}. Therefore, all partial derivatives of the function (3.1) relative to individual variables (*NPV* is an additive function of discounted revenues and costs) are greater than zero ($\partial NPV/\partial a_{el} > 0$, $\partial NPV/\partial a_{dust} > 0$, $\partial NPV/\partial a_{CO_2} > 0$ etc.), which means that the function (2.21) does not have an extreme, and changes in its value depend on from changes in value a_h, a_{el}, a_{fuel}, a_{CO_2}, a_{CO}, a_{SO_2}, a_{NO_X}, a_{dust}, b_{CO_2} and thus from changes in the time of price relations between energy carriers and environmental costs.

The equivalent criterion for maximizing profit $NPV \rightarrow \max$ when choosing the optimal technical solution for a power plant in dual-fuel gas-steam technology is to minimize the unit cost of electricity production in it $k_{el,av} \rightarrow \min$. With the condition $NPV = 0$ and assuming $a_{el} = 0$ (this assumption allows to calculate the average unit cost of electricity production $k_{el,av}$ during the T years of operation of the block), we obtain:

$$
\begin{aligned}
k_{el,av} = \frac{r}{1-\mathrm{e}^{-rT}} \Bigg\{ & \frac{E_{ch,A}^{gas}}{\left(E_{el,A}^{ST}+E_{el,A}^{GT}\right)(1-\varepsilon_{el})} \Bigg\{ (1+x_{sw,m,was}) e_{gas}^{t=0} \frac{1}{a_{gas}-r}[\mathrm{e}^{(a_{gas}-r)T}-1] \\
& + \rho_{CO_2}^{gas} p_{CO_2}^{t=0} \frac{1}{a_{CO_2}-r}[\mathrm{e}^{(a_{CO_2}-r)T}-1] + \rho_{CO}^{gas} p_{CO}^{t=0} \frac{1}{a_{CO}-r}[\mathrm{e}^{(a_{CO}-r)T}-1] \\
& + \rho_{NO_X}^{gas} p_{NO_X}^{t=0} \frac{1}{a_{NO_X}-r}[\mathrm{e}^{(a_{NO_X}-r)T}-1] + \rho_{SO_2}^{gas} p_{SO_2}^{t=0} \frac{1}{a_{SO_2}-r}[\mathrm{e}^{(a_{SO_2}-r)T}-1] \\
& + \rho_{dust}^{gas} p_{dust}^{t=0} \frac{1}{a_{dust}-r}[\mathrm{e}^{(a_{dust}-r)T}-1] + (1-u)\rho_{CO_2}^{gas} e_{CO_2}^{t=0} \frac{1}{b_{CO_2}-r}[\mathrm{e}^{(b_{CO_2}-r)T}-1] \Bigg\} \\
& + \frac{E_{ch,A}^{coal}}{\left(E_{el,A}^{ST}+E_{el,A}^{GT}\right)(1-\varepsilon_{el})} \Bigg\{ (1+x_{sw,m,was}) e_{coal}^{t=0} \frac{1}{a_{coal}-r}[\mathrm{e}^{(a_{coal}-r)T}-1] \\
& + \rho_{CO_2}^{coal} p_{CO_2}^{t=0} \frac{1}{a_{CO_2}-r}[\mathrm{e}^{(a_{CO_2}-r)T}-1] + \rho_{CO}^{coal} p_{CO}^{t=0} \frac{1}{a_{CO}-r}[\mathrm{e}^{(a_{CO}-r)T}-1] \\
& + \rho_{NO_X}^{coal} p_{NO_X}^{t=0} \frac{1}{a_{NO_X}-r}[\mathrm{e}^{(a_{NO_X}-r)T}-1] + \rho_{SO_2}^{coal} p_{SO_2}^{t=0} \frac{1}{a_{SO_2}-r}[\mathrm{e}^{(a_{SO_2}-r)T}-1] \\
& + \rho_{dust}^{coal} p_{dust}^{t=0} \frac{1}{a_{dust}-r}[\mathrm{e}^{(a_{dust}-r)T}-1] + (1-u)\rho_{CO_2}^{coal} e_{CO_2}^{t=0} \frac{1}{b_{CO_2}-r}[\mathrm{e}^{(b_{CO_2}-r)T}-1] \Bigg\} \\
& + \frac{J}{\left(E_{el,A}^{ST}+E_{el,A}^{GT}\right)(1-\varepsilon_{el})} (1-\mathrm{e}^{-rT})(1+x_{sal,t,ins}) \frac{\delta_{serv}}{r} \\
& + \frac{zJ}{\left(E_{el,A}^{ST}+E_{el,A}^{GT}\right)(1-\varepsilon_{el})} \left(\frac{1-\mathrm{e}^{-rT}}{T} + 1 \right) \Bigg\}.
\end{aligned}
\tag{3.2}
$$

3.1.1 *The Energy Balance of the Dual-Fuel Gas-Steam Power Plant in Series System*

Searching for the highest value of the target function (2.21), it is also possible to analyze the impact of the efficiency of devices used in particular technologies on its value. For this purpose, we should use the equations of state to express, for example, the amount $E_{ch,A}$ using energy balances. The energy balance of the dual-fuel gas-steam power plant in series is presented below (Fig. 3.1).

Using the above energy balance, for example, the gross efficiency of electricity generation in a dual-fuel gas-steam unit in a serial system is expressed in the formula (this efficiency can be about 45%) [1]:

$$
\eta_{el} = \frac{E_{el}^{GT} + E_{el}^{ST}}{E_{ch}^{gas} + E_{ch}^{coal}} = \frac{q_{ser}\eta_{GT} + [q_{ser}(1-\eta_{GT}) + 1]\eta_B \eta_{SH} \eta_{CR} \eta_i \eta_{me}}{1 + q_{ser}} \tag{3.3}
$$

where:

E_{ch}^{gas} chemical energy of gas burnt in a gas turbine,
E_{ch}^{coal} chemical energy of coal burned in the boiler,

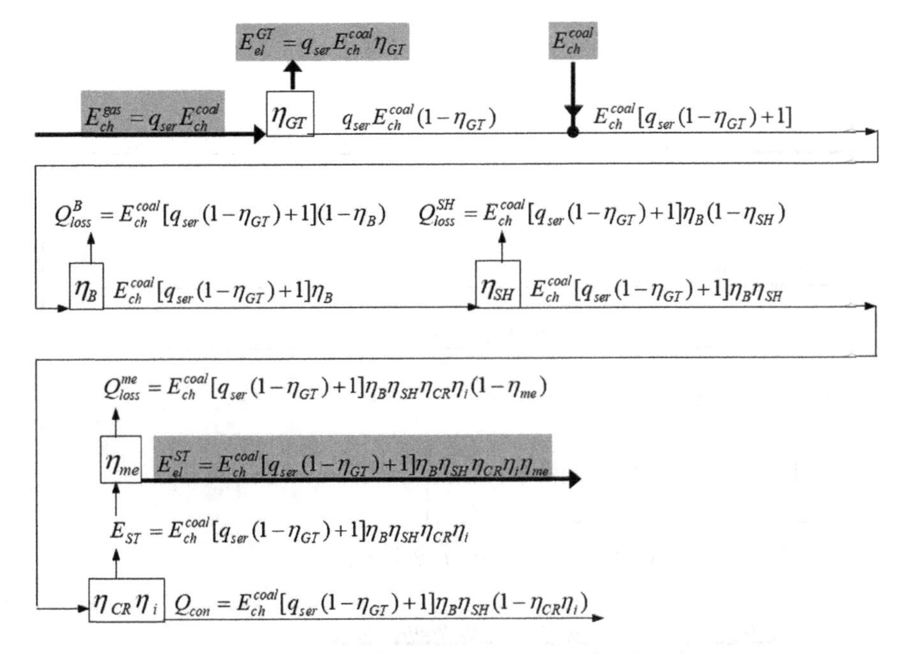

Fig. 3.1 The energy balance of the dual-fuel gas-steam power plant in series system

E_{el}^{GT}	gross electricity produced in a gas turboset,
E_{el}^{ST}	gross electricity produced in a steam turboset,
Q_{con}	heat of condensation in a steam turbine condenser,
q_{ser}	share of chemical energy of gas in chemical energy of coal in a serial system,
η_B	gross efficiency of the boiler,
η_{HRSG}	gross efficiency of a recovery boiler,
η_{SH}	energy efficiency of the collector system supplying steam to the turbine,
η_{GT}	gross energy efficiency of a gas turbine (energy efficiency of Joule's cycle, according to which a gas turbine works),
$\eta_{ST} = \eta_{CR}\eta_i$	energy efficiency of the steam turbine (the product of the energy efficiency of the Clausius-Rankine cycle at condensing operation and the internal efficiency of the steam turbine),
$\eta_{me} = \eta_m\eta_G$	electromechanical efficiency of a steam turbine set (efficiency product mechanical steam turbine and total generator efficiency).

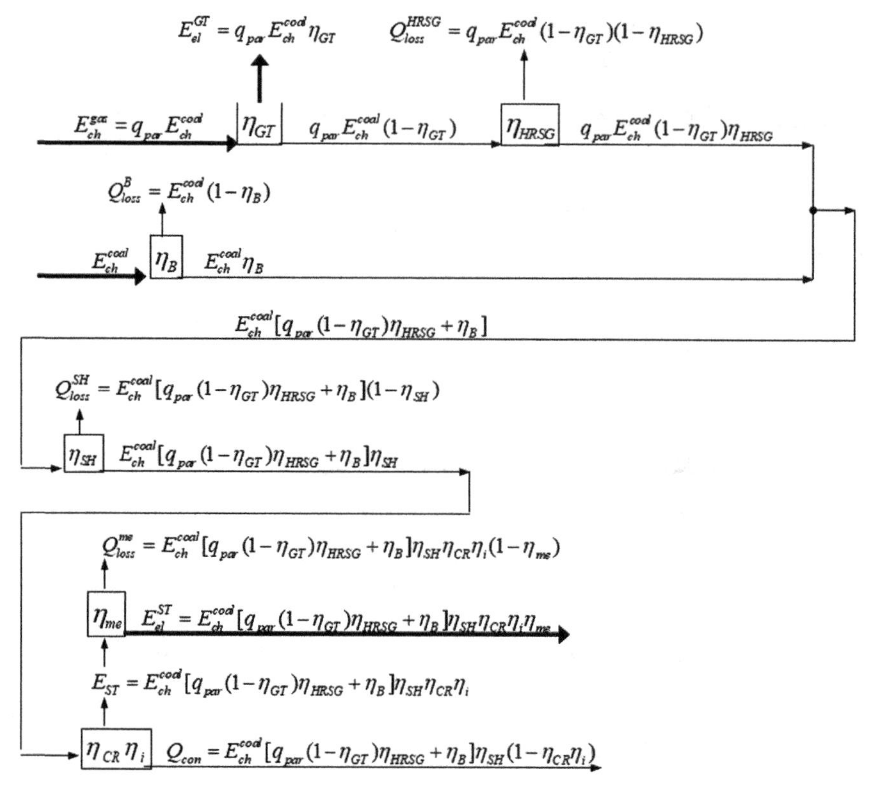

Fig. 3.2 The energy balance of a dual-fuel gas-steam power plant in a parallel system

3.1.2 *The Energy Balance of a Dual-Fuel Gas-Steam Power Plant in a Parallel System*

The balance of energy for a dual-fuel gas-steam power plant in a parallel system is shown below (Fig. 3.2).

where:

E_{el}^{ST}	gross electricity produced in a steam turboset,
E_{el}^{GT}	gross electricity generated in a gas turbine,
E_{ch}^{gas}	chemical energy of gas burnt in a gas turbine,
E_{ch}^{coal}	chemical energy of coal burned in the boiler,
Q_{con}	condensation heat in the condenser,
η_B	gross energy efficiency of the boiler,
η_{SH}	energy efficiency of the collector system supplying steam to the turbine,

$\eta_{ST} = \eta_{CR}\eta_i$ energy efficiency of the steam turbine circulation (the product of the energy efficiency of the *Clausius-Rankine* cycle and the internal efficiency of the steam turbine),

$\eta_{me} = \eta_m\eta_G$ electromechanical efficiency of the turbine set (the product of the mechanical efficiency of the steam turbine and the total efficiency of the generator),

η_{GT} gross energy efficiency of the Joule cycle of a gas turbine,

η_{HRSG} gross energy efficiency of a recovery boiler,

$q_{par} = E_{ch}^{gas}/E_{ch}^{coal}$ share of chemical energy of gas in chemical energy of coal combusted in a block.

3.2 Models for a Combined Heat and Power Plant

For a combined heat and power plant both in series as well as in parallel, regardless of whether they are equipped with steam and backpressure steam turbines or bleed-condensing steam turbines, the *NPV* gain is given by the following equation:

$$
\begin{aligned}
NPV = \Bigg\{ & (E_{el,A}^{ST} + E_{el,A}^{GT})(1-\varepsilon_{el})e_{el}^{t=0}\frac{1}{a_{el}-r}[e^{(a_{el}-r)T}-1] + Q_A e_c^{t=0}\frac{1}{a_c-r}[e^{(a_c-r)T}-1] \\
& - E_{ch,A}^{gas}\Bigg\{(1+x_{sw,m,was})e_{gas}^{t=0}\frac{1}{a_{gas}-r}[e^{(a_{gas}-r)T}-1] + \rho_{CO_2}^{gas}p_{CO_2}^{t=0}\frac{1}{a_{CO_2}-r}[e^{(a_{CO_2}-r)T}-1] \\
& + \rho_{CO}^{gas}p_{CO}^{t=0}\frac{1}{a_{CO}-r}[e^{(a_{CO}-r)T}-1] + \rho_{NO_X}^{gas}p_{NO_X}^{t=0}\frac{1}{a_{NO_X}-r}[e^{(a_{NO_X}-r)T}-1] \\
& + \rho_{SO_2}^{gas}p_{SO_2}^{t=0}\frac{1}{a_{SO_2}-r}[e^{(a_{SO_2}-r)T}-1] + \rho_{dust}^{gas}p_{dust}^{t=0}\frac{1}{a_{dust}-r}[e^{(a_{dust}-r)T}-1] \\
& + (1-u)\rho_{CO_2}^{gas}e_{CO_2}^{t=0}\frac{1}{b_{CO_2}-r}[e^{(b_{CO_2}-r)T}-1]\Bigg\} \\
& - E_{ch,A}^{coal}\Bigg\{(1+x_{sw,m,was})e_{coal}^{t=0}\frac{1}{a_{coal}-r}[e^{(a_{coal}-r)T}-1] + \rho_{CO_2}^{coal}p_{CO_2}^{t=0}\frac{1}{a_{CO_2}-r}[e^{(a_{CO_2}-r)T}-1] \\
& + \rho_{CO}^{coal}p_{CO}^{t=0}\frac{1}{a_{CO}-r}[e^{(a_{CO}-r)T}-1] + \rho_{NO_X}^{coal}p_{NO_X}^{t=0}\frac{1}{a_{NO_X}-r}[e^{(a_{NO_X}-r)T}-1] \\
& + \rho_{SO_2}^{coal}p_{SO_2}^{t=0}\frac{1}{a_{SO_2}-r}[e^{(a_{SO_2}-r)T}-1] + \rho_{dust}^{coal}p_{dust}^{t=0}\frac{1}{a_{dust}-r}[e^{(a_{dust}-r)T}-1] \\
& + (1-u)\rho_{CO_2}^{coal}e_{CO_2}^{t=0}\frac{1}{b_{CO_2}-r}[e^{(b_{CO_2}-r)T}-1]\Bigg\} \\
& - J(1-e^{-rT})(1+x_{sal,t,ins})\frac{\delta_{serv}}{r} - zJ\left(\frac{1-e^{-rT}}{T}+1\right)\Bigg\}(1-p)
\end{aligned}
\tag{3.4}
$$

where:

$a_{el}, a_{gas}, a_{coal}, a_{CO_2}, a_{CO}, a_{SO_2}, a_{NO_X}, a_{dust}, b_{CO_2}$–exponents expressing time evolution of the price of energy carriers and charges for the emission of harmful combustion products into the environment (e.g. $e_{el}(t) = e_{el}^{t=0}e^{a_{el}t}$, etc.; depending on the a_{el} value, the e_{el} price may increase, decrease or remain unchanged in subsequent years),

$e_{el}, e_{gas}, e_{coal}, e_{CO_2}$ unit price of electricity, natural gas, coal and purchase of CO_2 emission allowances,

$E_{el,A}^{ST}$ annual gross electricity production in a steam turboset,
$E_{el,A}^{GT}$ annual gross electricity production in a gas turbine,
$E_{ch,A}^{gas}$ annual consumption of chemical energy of gas burnt in a gas turbine,
$E_{ch,A}^{coal}$ annual consumption of chemical energy of coal burned in the boiler,
η_B gross energy efficiency of the boiler,
η_{SH} energy efficiency of the collector system supplying steam to the turbine,
$\eta_{ST} = \eta_{CR}\eta_i$ energy efficiency of the steam turbine circulation (the product of the energy efficiency of the Clausius-Rankine cycle and the internal efficiency of the steam turbine),
$\eta_{me} = \eta_m\eta_G$ electromechanical efficiency of the turbine set (the product of the mechanical efficiency of the steam turbine and the total efficiency of the generator),
η_{GT} gross energy efficiency of the Joule cycle of the gas turbine,
η_{HRSG} gross energy efficiency of a recovery boiler,
$p_{CO_2}, p_{CO}, p_{NO_x}, p_{SO_2}, p_{dust}$ charges for the emission of harmful combustion products into the environment,
p the income tax rate on gross profit,
Q_A annual production of heating,
r discount rate,
u share of chemical energy of fuel in its total annual consumption, for which it is not required to purchase CO_2 emission permits,
$x_{sw,m,was}$ factor taking into account the costs of supplementary water, auxiliary materials, sewage disposal, storage of slag, waste (in practice, the value of $x_{sw,m,was}$ is approx. 0.25),
$x_{sal,t,\,ns}$ factor including costs of wages, taxes, insurance, etc. (in practice, the value of $x_{sal,t,ins}$ is approx. 0.02),
z freezing capital investment ratio,
δ_{serv} the rate of fixed costs depending on investment expenditures (maintenance costs, overhauls of equipment),
ε_{el} indicator of electrical own needs of the power block,
$\rho_{CO_2}, \rho_{CO}, \rho_{NO_x}, \rho_{SO_2}\rho_{dust}$ CO_2, CO, NO_x, SO_2, and dust emissions per unit of chemical fuel energy.

From the formula (3.4) from the conditions $NPV = 0$ and $a_h = 0$, the average unit cost of heat production in a dual-fuel gas-steam combined heat and power plant is determined:

$$
\begin{aligned}
k_{h,av} = \frac{r}{1-\mathrm{e}^{-rT}} \Bigg\{ & \frac{E_{ch,A}^{gas}}{Q_A} \Bigg\{ (1 + x_{sw,m,was}) e_{gas}^{t=0} \frac{1}{a_{gas} - r} [\mathrm{e}^{(a_{gas}-r)T} - 1] \\
& + \rho_{\mathrm{CO_2}}^{gas} p_{\mathrm{CO_2}}^{t=0} \frac{1}{a_{\mathrm{CO_2}} - r} [\mathrm{e}^{(a_{\mathrm{CO_2}}-r)T} - 1] + \rho_{\mathrm{CO}}^{gas} p_{\mathrm{CO}}^{t=0} \frac{1}{a_{\mathrm{CO}} - r} [\mathrm{e}^{(a_{\mathrm{CO}}-r)T} - 1] \\
& + \rho_{\mathrm{NO_X}}^{gas} p_{\mathrm{NO_X}}^{t=0} \frac{1}{a_{\mathrm{NO_X}} - r} [\mathrm{e}^{(a_{\mathrm{NO_X}}-r)T} - 1] \\
& + \rho_{\mathrm{SO_2}}^{gas} p_{\mathrm{SO_2}}^{t=0} \frac{1}{a_{\mathrm{SO_2}} - r} [\mathrm{e}^{(a_{\mathrm{SO_2}}-r)T} - 1] + \rho_{dust}^{gas} p_{dust}^{t=0} \frac{1}{a_{dust} - r} [\mathrm{e}^{(a_{dust}-r)T} - 1] \\
& + (1-u) \rho_{\mathrm{CO_2}}^{gas} e_{\mathrm{CO_2}}^{t=0} \frac{1}{b_{\mathrm{CO_2}} - r} [\mathrm{e}^{(b_{\mathrm{CO_2}}-r)T} - 1] \Bigg\} \\
& + \frac{E_{ch,A}^{coal}}{Q_A} \Bigg\{ (1 + x_{sw,m,was}) e_{coal}^{t=0} \frac{1}{a_{coal} - r} [\mathrm{e}^{(a_{coal}-r)T} - 1] \\
& + \rho_{\mathrm{CO_2}}^{coal} p_{\mathrm{CO_2}}^{t=0} \frac{1}{a_{\mathrm{CO_2}} - r} [\mathrm{e}^{(a_{\mathrm{CO_2}}-r)T} - 1] \\
& + \rho_{\mathrm{CO}}^{coal} p_{\mathrm{CO}}^{t=0} \frac{1}{a_{\mathrm{CO}} - r} [\mathrm{e}^{(a_{\mathrm{CO}}-r)T} - 1] + \rho_{\mathrm{NO_X}}^{coal} p_{\mathrm{NO_X}}^{t=0} \frac{1}{a_{\mathrm{NO_X}} - r} [\mathrm{e}^{(a_{\mathrm{NO_X}}-r)T} - 1] \\
& + \rho_{\mathrm{SO_2}}^{coal} p_{\mathrm{SO_2}}^{t=0} \frac{1}{a_{\mathrm{SO_2}} - r} [\mathrm{e}^{(a_{\mathrm{SO_2}}-r)T} - 1] + \rho_{dust}^{coal} p_{dust}^{t=0} \frac{1}{a_{dust} - r} [\mathrm{e}^{(a_{dust}-r)T} - 1] \\
& + (1-u) \rho_{\mathrm{CO_2}}^{coal} e_{\mathrm{CO_2}}^{t=0} \frac{1}{b_{\mathrm{CO_2}} - r} [\mathrm{e}^{(b_{\mathrm{CO_2}}-r)T} - 1] \Bigg\} \\
& + \frac{J}{Q_A} (1 - \mathrm{e}^{-rT})(1 + x_{sal,t,ins}) \frac{\delta_{serv}}{r} + \frac{zJ}{Q_A} \left(\frac{1-\mathrm{e}^{-rT}}{T} + 1 \right) \\
& - \frac{E_{el,A}^{ST} + E_{el,A}^{GT}}{Q_A} (1 - \varepsilon_{el}) e_{el}^{t=0} \frac{1}{a_{el} - r} [\mathrm{e}^{(a_{el}-r)T} - 1] \Bigg\}.
\end{aligned}
\tag{3.5}
$$

3.2.1 Energy Balances of Dual-Fuel Gas-Steam Combined Heat and Power Plant in Series System

Below are presented the energy balances of a dual-fuel gas-steam combined heat and power plant in a series system. They are necessary, for example, in the analysis of the impact of energy efficiency of the devices used on the value of *NPV* generated from the operation of the combined heat and power plant.

3.2.1.1 A System with a Bleed-Condensing Steam Turbine

Below is presented the energy balance of a dual-fuel gas-steam combined heat and power plant in a series system with a bleed-condensing steam turbine (Fig. 3.3).

Whereby the parameter β means the ratio of heat contained in the bleeding heating steam supplying the *HE* heat exchangers to the mechanical energy generated in the bleed-condensing steam turbine: $\beta = Q_u/E_{ST}^{k\ \mathrm{mod}}$. The parameter β is the inverse of the association index σ of the steam turbine.

For example, the gross efficiency of heat and electricity generation in a combined heat and power plant (Hot Windbox) with a bleed-condensing steam turbine is expressed by the formula:

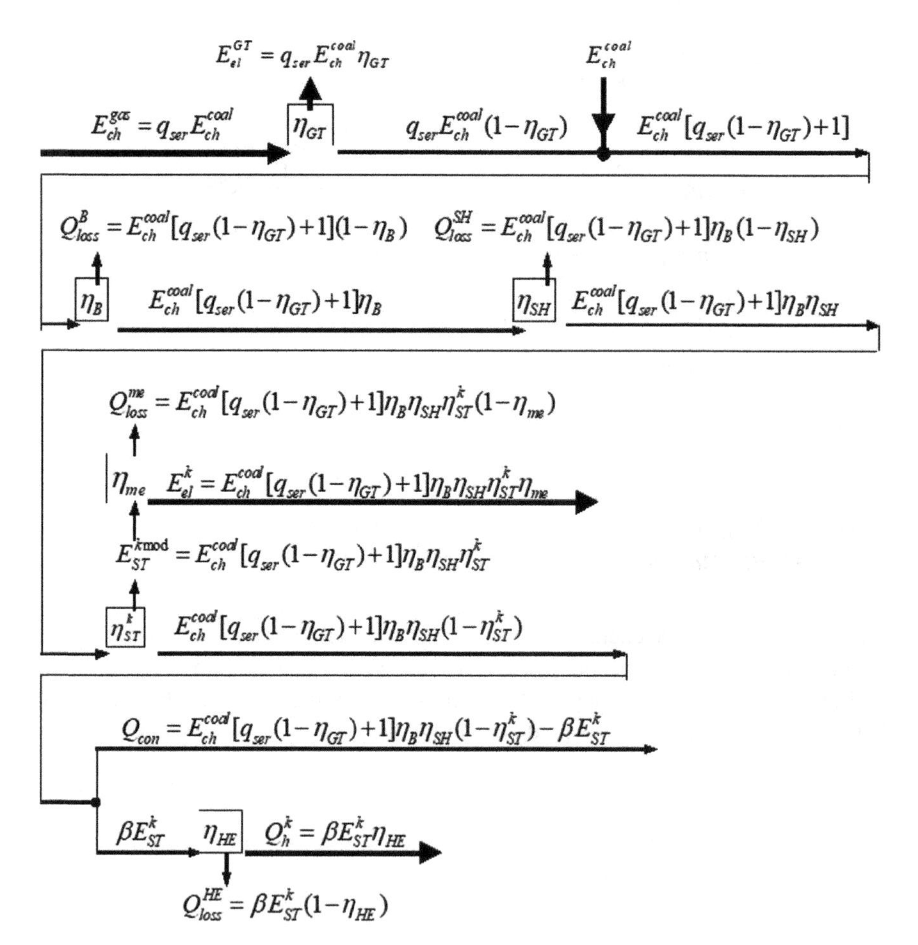

Fig. 3.3 Energy balance of a dual-fuel gas-steam combined heat and power plant in a series system with a bleed-condensing steam turbine

$$\eta_h = \frac{E_{el}^{GT} + E_{el}^{k} + Q_h^k}{E_{ch}^{gas} + E_{ch,w}} = \frac{q_{ser}\eta_{GT} + \{[q_{ser}(1-\eta_{GT}) + 1]\eta_B\eta_{SH}\eta_{CR}\eta_i\}(\eta_{me} + \beta\eta_{HE})}{1 + q_{ser}} \tag{3.6}$$

where:

E_{el}^{k}	gross electricity produced in a bleed-condensing steam turboset in a combined heat and power plant,
E_{el}^{GT}	gross electricity generated in a gas turbine,
E_{ch}^{gas}	chemical energy of gas burnt in a gas turbine,
E_{ch}^{coal}	chemical energy of coal burned in the boiler,
$q_{ser} = E_{ch}^{gas}/E_{ch}^{coal}$	share of chemical energy of gas in a series system in the chemical energy of coal combusted in a modernized combined heat and power plant,
Q_h^k	gross heating produced in the combined heat and power plant in the condensing system of its operation,
Q_{con}	heat of steam condensation in bleed-condensing steam turbine condenser,
η_B	efficiency of the boiler,
η_{SH}	energy efficiency of the collector system supplying steam to the turbine,
$\eta_{ST}^k = \eta_{CR}^k\eta_i^k$	energy efficiency of the condensation steam turbine (the product of the energy efficiency of the Clausius-Rankine cycle and the internal efficiency of the steam turbine),
$\eta_{me} = \eta_m\eta_G$	electromechanical efficiency of the turbine set (product of mechanical efficiency the steam turbine and the efficiency of the total generator; it was assumed that electromechanical efficiency of backpressure and condensing turbines are the same),
η_{GT}	energy efficiency of a gas turbine,
η_{HE}	energy efficiency of heat exchangers station.

3.2.1.2 The System with the Bleed-Backpressure Steam Turbine

On figure below (Fig. 3.4) the balance sheet of a dual-fuel gas-steam combined heat and power plant in series with a backpressure steam turbine is presented.

E_{el}^{p}	gross electricity generated in a steam-backpressure turboset in a combined heat and power plant,
E_{el}^{GT}	gross electricity generated in a gas turbine,
E_{ch}^{gas}	chemical energy of gas burnt in a gas turbine,
E_{ch}^{coal}	chemical energy of coal burned in boilers,
$q_{ser} = E_{ch}^{gas}/E_{ch}^{coal}$	share of chemical energy of gas in a series system in the chemical energy of coal combusted in a modernized combined heat and power plant,
Q_h^p	gross heating produced in a combined heat and power plant with a backpressure steam turbine,

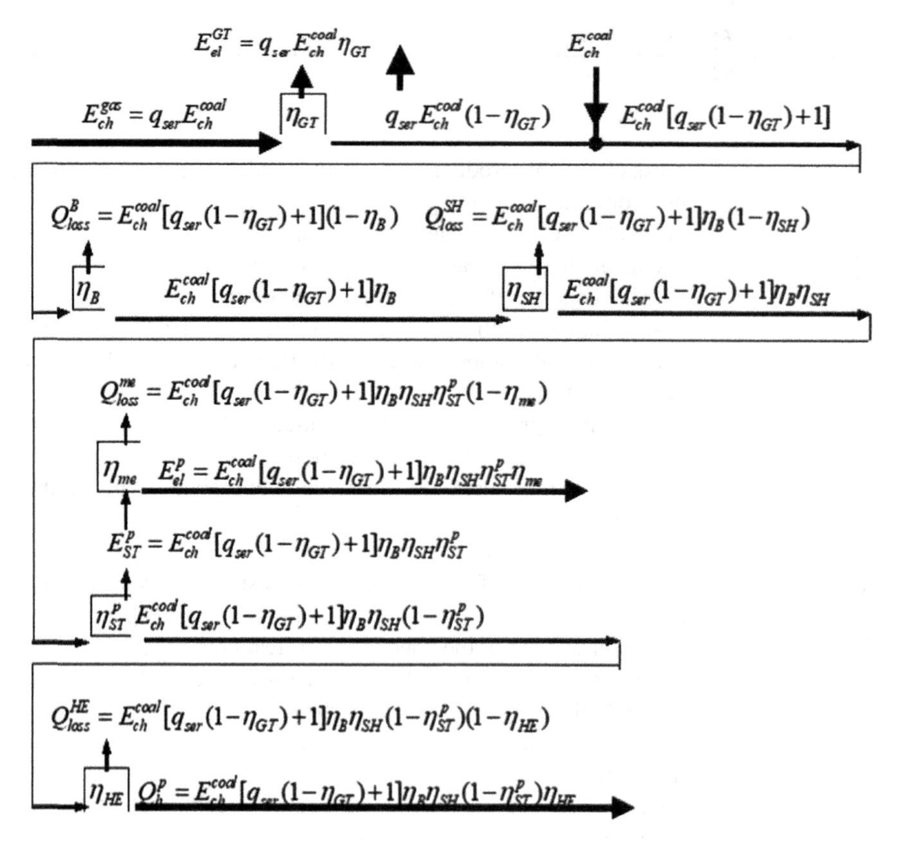

Fig. 3.4 Energy balance of a dual-fuel gas-steam combined heat and power plant in a series system with a backpressure steam turbine

η_B	gross efficiency of the boiler,
η_{SH}	energy efficiency of the collector system supplying steam to the turbine,
$\eta_{ST}^{p} = \eta_{CR}^{p}\eta_{i}^{p}$	energy efficiency of the backpressure-steam turbine (the product of the energy efficiency of the Clausius-Rankine cycle and the internal efficiency of the steam turbine),
$\eta_{me} = \eta_m \eta_G$	electromechanical efficiency of the turbine set (product of mechanical efficiency of the steam turbine and the total efficiency of the generator; it was assumed that electromechanical efficiency of backpressure and condensing turbines are the same),
η_{GT}	energy efficiency of a gas turbine,
η_{HE}	energy efficiency of heat exchangers station.

3.2.2 *Energy Balances of a Dual-Fuel Gas-Steam Combined Heat and Power Plant in a Parallel System*

3.2.2.1 A System with a Bleed-Condensing Steam Turbine

See Fig. 3.5.

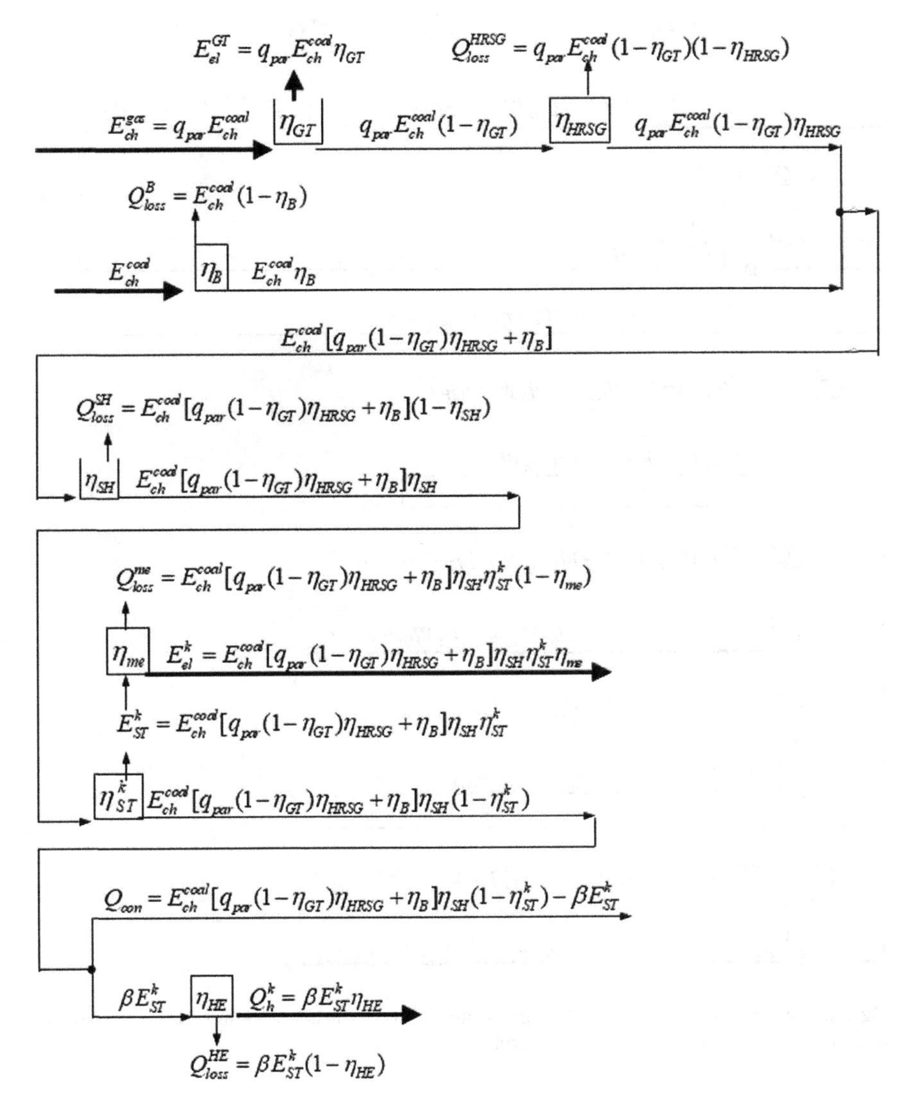

Fig. 3.5 Energy balance of a dual-fuel gas-steam combined heat and power plant in a parallel system with a steam-extraction and condensing steam turbine

3.2.2.2 A System with a Bleed-Backpressure Turbine

See Fig. 3.6.

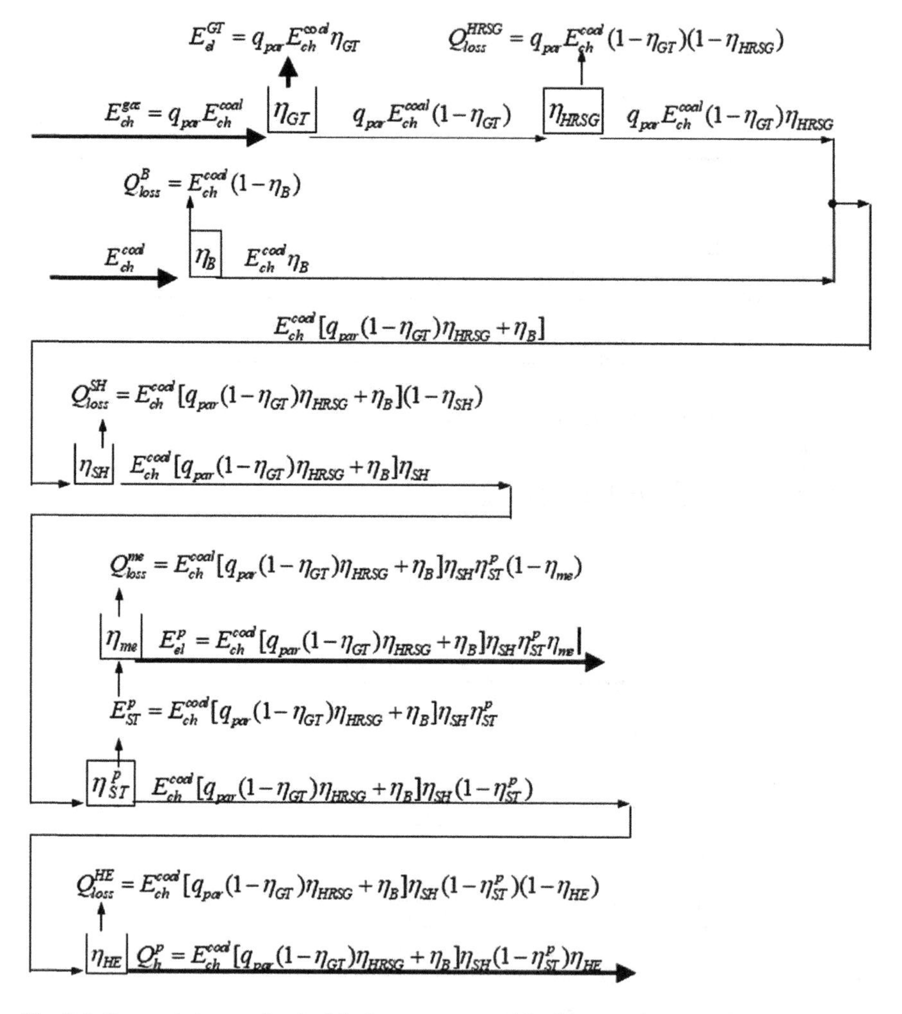

Fig. 3.6 Energy balance of a dual-fuel gas-steam combined heat and power plant in a parallel system with a bleed-backpressure steam turbine

References

1. Bartnik R, Bartnik B, Hnydiuk-Stefan A (2016) Optimum investment strategy in the power industry mathematical models. Springer, New York
2. Bartnik R, Buryn Z, Hnydiuk-Stefan A (2017) Investment strategy in heating and CHP mathematical models. Wydawnictwo. Springer, London

Chapter 4
Mathematical Models with the Continuous Time for Selection of the Optimum Power of a Gas Turbine Set for Newly Built Dual-Fuel Gas-Fired Combined Heat and Power Plants in Parallel Systems

In the market economy, the cost-effectiveness of building new power units and modernizing existing ones is the economic viability of their work. Precisely speaking, as already indicated in Chap. 2, the maximization of the achieved profit:

$$NPV \rightarrow \max. \tag{4.1}$$

The economic criterion is superior to the technical criterion. Technical analysis, although very important and necessary, should be used only when searching for opportunities to improve technological and technical processes, while improving structural solutions of machines and devices. Ultimately, however, this economic criterion, the criterion of profit maximization determines the purpose and selection of a specific technical solution.

The equivalent criterion for maximizing profit $NPV \rightarrow \max$ in the case of a power plant is to minimize the unit cost of electricity production in it

$$k_{el,av} \rightarrow \min, \tag{4.2}$$

and in the case of combined heat and power plants, minimization of the unit cost of heat production in it

$$k_{h,av} \rightarrow \min. \tag{4.3}$$

A. Hnydiuk-Stefan, *Dual-Fuel Gas-Steam Power Block Analysis*, Power Systems,
https://doi.org/10.1007/978-3-030-03050-6_4

4.1 Selection of the Optimal Power of the Gas Turbine Set for the Newly Built Dual-Fuel Gas-Steam Power Plant in a Parallel System

In the bi-fuel system, its steam part, boiler and steam turbine set can be built on supercritical parameters of fresh steam or subcritical parameters. Subcritical parameters do not cause a significant reduction in the gross efficiency of electricity generation in a dual-fuel gas-steam unit in a parallel system. This efficiency is expressed by the formula:

$$\eta_{el}^{GP,dp} = \frac{E_{el}^{GT} + E_{el}^{ST}}{E_{ch}^{gas} + E_{ch}^{coal}} = \frac{q_{par}\eta_{GT} + [q_{par}(1-\eta_{GT})\eta_{HRSG} + \eta_{B}]\eta_{SH}\eta_{ST}\eta_{me}}{1 + q_{par}}. \quad (4.4)$$

The above formula was obtained using the energy balance shown in Fig. 3.2.

The advantage of such parameters is, however, the significantly lower unit (per unit of power) investment outlay on the steam part of the system, which is approximately $i^{ST} = 4$ mln PLN/MW, when the supercritical parameters are approximately $i^{ST} \cong 6.5$ mln PLN/MW. The supercritical parameters therefore significantly increase the capital expenditures on the dual-fuel gas-steam system, but thanks to them, however, the efficiency of electricity generation in them is greater, hence the emission of pollutants to the environment is lower, e.g. carbon dioxide (Fig. 4.5). In this monograph two calculations were carried out, once assuming that $i^{ST} \cong 6.5$ mln PLN/MW and $\eta_{ST} = 0.55$ and second that $i^{ST} = 4$ mln PLN/MW and $\eta_{ST} = 0.45$. This allows to find the answer to the following questions:

(1) Which parameters are more cost-effective: supercritical or subcritical?
(2) What is optimal power of the gas turbine set in dual-fuel gas systems for both thermal parameters of the fresh steam, depending on the price relations between energy carriers and on investment outlays and environmental charges, especially on the purchase price of carbon dioxide emission permits?

When the (4.5), (4.6) and dependence (4.7) obtained from the balance of energy presented in Fig. 3.2 is based on (3.2), we get:

$$\frac{E_{ch,A}^{gas}}{E_{el,A}^{ST} + E_{el,A}^{GT}} = \frac{q_{par}}{[q_{par}(1-\eta_{GT})\eta_{HRSG} + \eta_{B}]\eta_{SH}\eta_{ST}\eta_{me} + q_{par}\eta_{GT}} \quad (4.5)$$

$$\frac{E_{ch,A}^{coal}}{E_{el,A}^{ST} + E_{el,A}^{GT}} = \frac{1}{[q_{par}(1-\eta_{GT})\eta_{HRSG} + \eta_{B}]\eta_{SH}\eta_{ST}\eta_{me} + q_{par}\eta_{GT}} \quad (4.6)$$

$$\frac{J}{\left(E_{el,A}^{ST}+E_{el,A}^{GT}\right)(1-\varepsilon_{el})}=\frac{\left(N_{el}^{ST}+N_{el}^{GT}\right)i}{\left(N_{el}^{ST}+N_{el}^{GT}\right)\tau_A(1-\varepsilon_{el})} \tag{4.7}$$

where:

N_{el}^{ST}, N_{el}^{GT} gross power of the steam and gas turbine set,
τ_A annual block operation time,

The desired form of the formula is obtained for the average unit cost of $k_{el,av}$, i.e. the form in which the independent variable q_{par} exists:

$$\begin{aligned}
k_{el,av} = {} & \frac{r}{(1-\mathrm{e}^{-rT})(1-\varepsilon_{el})}\left\{\frac{q_{par}}{[q_{par}(1-\eta_{GT})\eta_{HRSG}+\eta_B]\eta_{col}\eta_{ST}\eta_{me}+q_{par}\eta_{GT}}\right. \\
& \times\left\{(1+x_{sw,m,was})e_{gas}^{t=0}\frac{1}{a_{gas}-r}[\mathrm{e}^{(a_{gas}-r)T}-1]\right. \\
& +\rho_{CO_2}^{gas}p_{CO_2}^{t=0}\frac{1}{a_{CO_2}-r}[\mathrm{e}^{(a_{CO_2}-r)T}-1]+\rho_{CO}^{gas}p_{CO}^{t=0}\frac{1}{a_{CO}-r}[\mathrm{e}^{(a_{CO}-r)T}-1] \\
& +\rho_{NO_X}^{gas}p_{NO_X}^{t=0}\frac{1}{a_{NO_X}-r}[\mathrm{e}^{(a_{NO_X}-r)T}-1]+\rho_{SO_2}^{gas}p_{SO_2}^{t=0}\frac{1}{a_{SO_2}-r}[\mathrm{e}^{(a_{SO_2}-r)T}-1] \\
& \left.+\rho_{dust}^{gas}p_{dust}^{t=0}\frac{1}{a_{dust}-r}[\mathrm{e}^{(a_{dust}-r)T}-1]+(1-u)\rho_{CO_2}^{gas}e_{CO_2}^{t=0}\frac{1}{b_{CO_2}-r}[\mathrm{e}^{(b_{CO_2}-r)T}-1]\right\} \\
& +\frac{1}{[q_{par}(1-\eta_{GT})\eta_{HRSG}+\eta_B]\eta_{col}\eta_{ST}\eta_{me}+q_{par}\eta_{GT}} \\
& \times\left\{(1+x_{sw,m,was})e_{coal}^{t=0}\frac{1}{a_{coal}-r}[\mathrm{e}^{(a_{coal}-r)T}-1]\right. \\
& +\rho_{CO_2}^{coal}p_{CO_2}^{t=0}\frac{1}{a_{CO_2}-r}[\mathrm{e}^{(a_{CO_2}-r)T}-1]+\rho_{CO}^{coal}p_{CO}^{t=0}\frac{1}{a_{CO}-r}[\mathrm{e}^{(a_{CO}-r)T}-1] \\
& +\rho_{NO_X}^{coal}p_{NO_X}^{t=0}\frac{1}{a_{NO_X}-r}[\mathrm{e}^{(a_{NO_X}-r)T}-1]+\rho_{SO_2}^{coal}p_{SO_2}^{t=0}\frac{1}{a_{SO_2}-r}[\mathrm{e}^{(a_{SO_2}-r)T}-1] \\
& \left.+\rho_{dust}^{coal}p_{dust}^{t=0}\frac{1}{a_{dust}-r}[\mathrm{e}^{(a_{dust}-r)T}-1]+(1-u)\rho_{CO_2}^{coal}e_{CO_2}^{t=0}\frac{1}{b_{CO_2}-r}[\mathrm{e}^{(b_{CO_2}-r)T}-1]\right\} \\
& \left.+\frac{i}{\tau_A}\left[(1-\mathrm{e}^{-rT})(1+x_{sal,t,ins})\frac{\delta_{serv}}{r}+z\left(\frac{1-\mathrm{e}^{-rT}}{T}+1\right)\right]\right\}.
\end{aligned} \tag{4.8}$$

In the formula (4.8) also unit outlays i are a function of the value q_{par}. But, without making a significant mistake, it can be assumed that their value is constant, i = const, so the J outlays depend only on the total rated electrical power of the plant $J = (N_{el}^{ST} + N_{el}^{GT})i$ formula (4.7). It should be remembered that the unit expenditures "i" are a function of both unit expenditures on the gas part of the block i^{GT}, and on the steam part i^{ST} : $i = f(i^{GT}, i^{ST})$.

What is more, the unit outlays i^{GT}, i^{ST} are the smaller the larger the power of the gas N_{el}^{GT} and steam turbine set N_{el}^{ST}. Moreover, unit expenditures i^{GT} are significantly smaller than i^{ST} expenditures. Thus, for the current price relations between energy carriers, the powers N_{el}^{GT} and N_{el}^{ST} have already been determined using formula (4.12), then in calculations checking the economic profitability of the block's operation, investment outlays should be calculated from the relation $J = N_{el}^{ST} i^{ST} + N_{el}^{GT} i^{GT}$ for the determined powers N_{el}^{ST} and N_{el}^{GT} for i^{ST}, i^{GT} values.

As already indicated above, the monograph present the results of calculations for two variants of investment outlays, once for $i \cong 4.6$ mln PLN/MW, that is when the steam part works with supercritical parameters of a fresh steam, and two when $i \cong 3.4$ mln PLN/MW, so for subcritical parameters. In the first case, the efficiency of the *Clausius-Rankine* cycle was equal to $\eta_{ST} = 0.55$, in the second equal to $\eta_{ST} = 0.45$.

Formula (4.8), allows for a wide analysis of the impact of an independent variable q_{par}, as well as price relations between the price of gas and coal and the value of energy efficiency of devices used in gas-steam systems for the optimal value of the power of the gas turbine set.

When calculating from the formula (4.8) the derivative $dk_{el,av} / dq_{par}$ we obtain:

$$\begin{aligned}
\frac{dk_{el,av}}{dq_{par}} = {} & \frac{r}{(1-e^{-rT})(1-\varepsilon_{el})}\Bigg\{\frac{\eta_B \eta_{col} \eta_{ST} \eta_{me}}{\{[q_{par}(1-\eta_{GT})\eta_{HRSG}+\eta_B]\eta_{col}\eta_{ST}\eta_{me}+q_{par}\eta_{GT}\}^2} \\
& \times \Bigg\{(1+x_{sw,m,was}) e_{gas}^{t=0} \frac{1}{a_{gas}-r}[e^{(a_{gas}-r)T}-1] \\
& + \rho_{CO_2}^{gas} p_{CO_2}^{t=0} \frac{1}{a_{CO_2}-r}[e^{(a_{CO_2}-r)T}-1] + \rho_{CO}^{gas} p_{CO}^{t=0} \frac{1}{a_{CO}-r}[e^{(a_{CO}-r)T}-1] \\
& + \rho_{NO_X}^{gas} p_{NO_X}^{t=0} \frac{1}{a_{NO_X}-r}[e^{(a_{NO_X}-r)T}-1] + \rho_{SO_2}^{gas} p_{SO_2}^{t=0} \frac{1}{a_{SO_2}-r}[e^{(a_{SO_2}-r)T}-1] \\
& + \rho_{dust}^{gas} p_{dust}^{t=0} \frac{1}{a_{dust}-r}[e^{(a_{dust}-r)T}-1] + (1-u)\rho_{CO_2}^{gas} e_{CO_2}^{t=0} \frac{1}{b_{CO_2}-r}[e^{(b_{CO_2}-r)T}-1]\Bigg\} \\
& - \frac{(1-\eta_{GT})\eta_{HRSG}\eta_{col}\eta_{ST}\eta_{me}+\eta_{GT}}{\{[q_{par}(1-\eta_{GT})\eta_{HRSG}+\eta_B]\eta_{col}\eta_{ST}\eta_{me}+q_{par}\eta_{GT}\}^2} \\
& \times \Bigg\{(1+x_{sw,m,was}) e_{coal}^{t=0} \frac{1}{a_{coal}-r}[e^{(a_{coal}-r)T}-1] \\
& + \rho_{CO_2}^{coal} p_{CO_2}^{t=0} \frac{1}{a_{CO_2}-r}[e^{(a_{CO_2}-r)T}-1] + \rho_{CO}^{coal} p_{CO}^{t=0} \frac{1}{a_{CO}-r}[e^{(a_{CO}-r)T}-1] \\
& + \rho_{NO_X}^{coal} p_{NO_X}^{t=0} \frac{1}{a_{NO_X}-r}[e^{(a_{NO_X}-r)T}-1] + \rho_{SO_2}^{coal} p_{SO_2}^{t=0} \frac{1}{a_{SO_2}-r}[e^{(a_{SO_2}-r)T}-1] \\
& + \rho_{dust}^{coal} p_{dust}^{t=0} \frac{1}{a_{dust}-r}[e^{(a_{dust}-r)T}-1] + (1-u)\rho_{CO_2}^{coal} e_{CO_2}^{t=0} \frac{1}{b_{CO_2}-r}[e^{(b_{CO_2}-r)T}-1]\Bigg\}\Bigg\}.
\end{aligned} \tag{4.9}$$

From the condition $dk_{el,av} / dq_{par} = 0$ a relation is obtain:

$$
\begin{aligned}
&\eta_B \eta_{col} \eta_{ST} \eta_{me} \left\{ (1 + x_{sw,m,was}) e_{gas}^{t=0} \frac{1}{a_{gas} - r} [\mathrm{e}^{(a_{gas} - r)T} - 1] \right. \\
&+ \rho_{CO_2}^{gzs} p_{CO_2}^{t=0} \frac{1}{a_{CO_2} - r} [\mathrm{e}^{(a_{CO_2} - r)T} - 1] + \rho_{CO}^{gas} p_{CO}^{t=0} \frac{1}{a_{CO} - r} [\mathrm{e}^{(a_{CO} - r)T} - 1] \\
&+ \rho_{NO_X}^{gas} p_{NO_X}^{t=0} \frac{1}{a_{NO_X} - r} [\mathrm{e}^{(a_{NO_X} - r)T} - 1] + \rho_{SO_2}^{gas} p_{SO_2}^{t=0} \frac{1}{a_{SO_2} - r} [\mathrm{e}^{(a_{SO_2} - r)T} - 1] \\
&\left. + \rho_{dust}^{gas} p_{dust}^{t=0} \frac{1}{a_{dust} - r} [\mathrm{e}^{(a_{dust} - r)T} - 1] + (1 - u) \rho_{CO_2}^{gas} e_{CO_2}^{t=0} \frac{1}{b_{CO_2} - r} [\mathrm{e}^{(b_{CO_2} - r)T} - 1] \right\} \\
&- [(1 - \eta_{GT}) \eta_{HRSG} \eta_{col} \eta_{ST} \eta_{me} + \eta_{GT}] \left\{ (1 + x_{sw,m,was}) e_{coal}^{t=0} \frac{1}{a_{coal} - r} [\mathrm{e}^{(a_{coal} - r)T} - 1] \right. \\
&+ \rho_{CO_2}^{coal} p_{CO_2}^{t=0} \frac{1}{a_{CO_2} - r} [\mathrm{e}^{(a_{CO_2} - r)T} - 1] + \rho_{CO}^{coal} p_{CO}^{t=0} \frac{1}{a_{CO} - r} [\mathrm{e}^{(a_{CO} - r)T} - 1] \\
&+ \rho_{NO_X}^{coal} p_{NO_X}^{t=0} \frac{1}{a_{NO_X} - r} [\mathrm{e}^{(a_{NO_X} - r)T} - 1] + \rho_{SO_2}^{coal} p_{SO_2}^{t=0} \frac{1}{a_{SO_2} - r} [\mathrm{e}^{(a_{SO_2} - r)T} - 1] \\
&\left. + \rho_{dust}^{coal} p_{dust}^{t=0} \frac{1}{a_{dust} - r} [\mathrm{e}^{(a_{dust} - r)T} - 1] + (1 - u) \rho_{CO_2}^{coal} e_{CO_2}^{t=0} \frac{1}{b_{CO_2} - r} [\mathrm{e}^{(b_{CO_2} - r)T} - 1] \right\} = 0
\end{aligned}
\quad (4.10)
$$

from which it follows that the amount q_{par} that limits (minimize or maximize) the cost $k_{el,av}$ (formula (4.8)) does not exist. The value q_{par} in the formula (4.10) is not present (this situation can change the dependence of unit outlays i on q_{par} value). From the formula (4.10), it is possible to determine, what is important, the gas e_{gas}^{bo} and coal e_{coal}^{bo} boundary prices, i.e. prices, for which the cost $k_{el,av}$ assumes a constant value, and therefore independent of q_{par}, because then the derivative fulfills the condition $dk_{el,av} / dq_{par} = 0$.

Thus, for a given coal price from the Eq. (4.10), a gas boundary price is set e_{gas}^{bo}, and vice versa, for a given gas price a boundary price of coal is set e_{coal}^{bo}—Fig. 4.4. Both of these prices are therefore closely related to each other, when substituted for Eq. (4.9) "reset" its value. On the other hand, if other values are inserted into Eq. (4.9), for example current coal and gas prices, different from boundary values, then the equation makes a positive or negative value.

If the costs associated with gas in Eq. (4.10) exceed the costs associated with coal, then $dk_{el,av} / dq_{par} > 0$ and cost rises with increase q_{par}, and vice versa, when the costs associated with gas are lower than the costs associated with coal, then the relation $dk_{el,av} / dq_{par} < 0$ is met and the cost $k_{el,av}$ decreases with the increase in power gas turbine set.

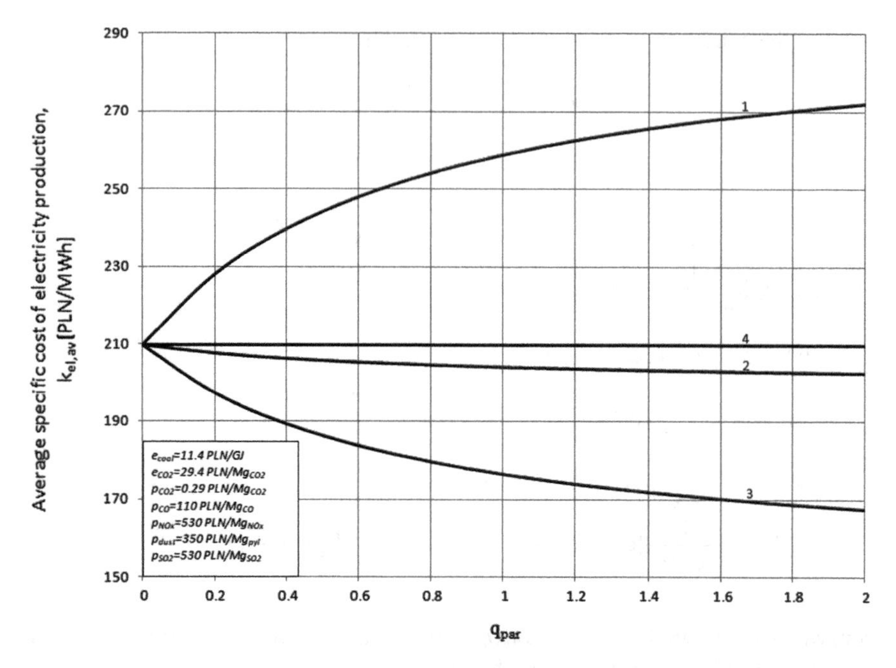

Fig. 4.1 Average specific cost of electricity production q_{par} in function with the price of gas as a parameter, where: 1 applies to $e_{gas} = 32$ PLN/GJ, $i = 4.6$ mln/MW, $\eta_{ST} = 0.55$; 2 applies to $e_{gas} = 16$ PLN/GJ, $i = 4.6$ mln/MW, $\eta_{ST} = 0.55$; 3 applies to $e_{gas} = 8$ PLN/GJ; $i = 4.6$ mln/MW, $\eta_{ST} = 0.55$; 4 applies to $e^{bo}_{gas} = 17.67$ PLN/GJ; $i = 4.6$ mln/MW, $\eta_{ST} = 0.55$

Therefore, the *necessary condition* for a dual-fuel gas-steam power block to be economically more profitable than a single coal block (i.e. when $q_{par} = 0$) is to meet the relation $dk_{el,av} / dq_{par} < 0$, i.e. that the gas price is lower than the boundary price designated for the current price of coal.

However, it should be noted that the fulfill relation $dk_{el,av} / dq_{par} < 0$ is possible in two cases. Once, when the price of coal is higher than its boundary price (then the cost $k_{el,av}$ is the highest, curves 5 in Figs. 4.2 and 4.2a) and two, when the price of gas is lower than its boundary price (then the cost $k_{el,av}$ is the lowest, curves 2, 3 in Figs. 4.1 and 4.1a).

Therefore, the *necessary condition* for a dual-fuel gas-steam power plant to be economically more profitable than a single coal-fired power plant is not only to meet the relation $dk_{el,av} / dq_{par} < 0$, but also that the gas price is lower than its boundary price determined from the Eq. (4.20) for the current price of coal and electricity (curves 2, 3 in Figs. 4.1 and 4.1a).

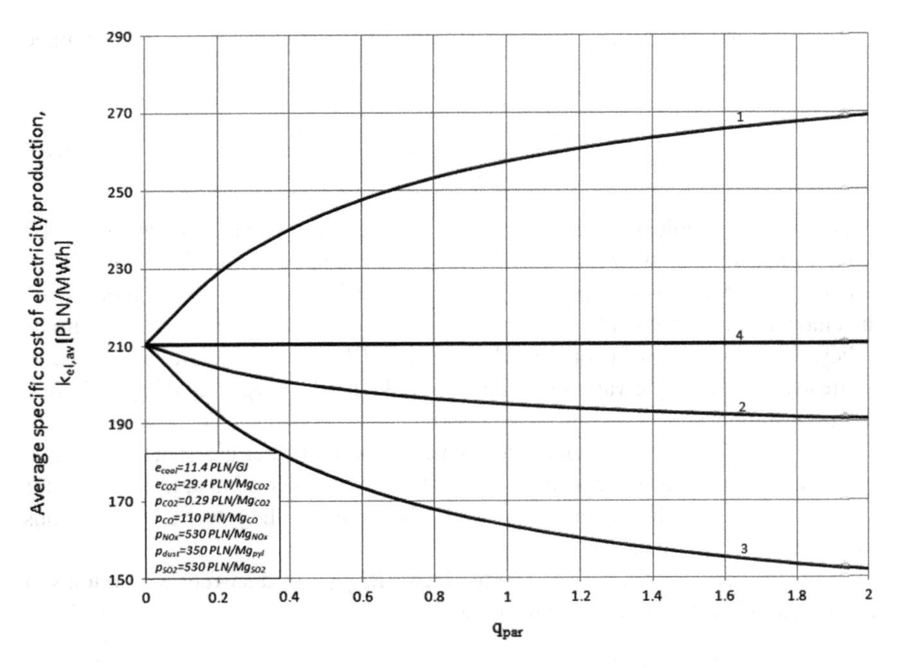

Fig. 4.1a Average specific cost of electricity production q_{par} in function with the price of gas as a parameter, where: 1 applies to e_{gas} = 32 PLN/GJ; i = 3.4 mln/MW, η_{ST} = 0.45; 2 applies to e_{gas} = 16 PLN/GJ; i = 3.4 mln/MW, η_{ST} = 0.45; 3 applies to e_{gas} = 8 PLN/GJ, i = 3.4 mln/MW, η_{ST} = 0.45; 4 applies to e^{bo}_{gas} = 20.01 PLN/GJ, i = 3.4 mln/MW, η_{ST} = 0.45

The *necessary condition* for the profitability of the construction of dual-fuel power blocks should therefore be finally recorded with the following relation:

$$e_{gas} \leq e^{bo}_{gas}. \tag{4.11}$$

The most profitable then is the largest power of the gas turbine set, and therefore the greatest $q_{par} = q^{\max}_{par}$ value. Knowledge of boundary prices e^{bo}_{coal}, e^{bo}_{gas} coal and gas is therefore very important.

By knowing them, it is possible to answer the question: is the dual-fuel gas-steam combined cycle more economically reasonable for the current prices of coal, gas, costs of emission of harmful products of the block's operation to the environment and investment outlays for its construction (capital expenditures decide on η_{ST} value) or more profitable is the single coal block?

The most favorable value of the power of the gas turbine set $N_{el\max}^{GT}$ is determined from the relation:

$$q_{par}^{\max} = \frac{E_{ch,}^{gas}}{E_{ch,coal}} \cong \frac{N_{el\max}^{GT}}{N_{el}^{ST}}. \tag{4.12}$$

An important indicator characterizing the operation of power units is the carbon dioxide emission factor EF_{CO_2}. This factor express the number of kilograms of CO_2 emitted per megawatt-hour produced E_{el} net electric energy in power block from the amount E_{ch} of chemical energy of the burned fuel. According to the EU climate policy, the value of this factor should be at most $EF_{CO_2} = 750\,\mathrm{kg_{CO_2}/MWh}$ (but it is already said about the values of the factor at the level $EF_{CO_2} = 500\,\mathrm{kg_{CO_2}/MWh}$, and even smaller).

As it has already been mentioned above, in the dual-fuel gas-steam combined cycle this factor, what is important, is significantly smaller compared to the factor for the block in which only coal is burned. This is due to the lower CO_2 emissions from natural gas combustion.

Using the relations (4.5) and (4.6) the factor EF_{CO_2} for dual-fuel gas systems in q_{par} function is expressed by the formula:

$$\begin{aligned} EF_{CO_2} &= \frac{E_{ch}^{coal}\rho_{CO_2}^{coal} + E_{ch}^{gas}\rho_{CO_2}^{gas}}{\left(E_{el}^{ST} + E_{el}^{GT}\right)(1-\varepsilon_{el})} \\ &= \frac{\rho_{CO_2}^{coal} + q_{par}\rho_{CO_2}^{gas}}{\left\{[q_{par}(1-\eta_{GT})\eta_{HRSG} + \eta_B]\eta_{col}\eta_{ST}\eta_{me} + q_{par}\eta_{GT}\right\}(1-\varepsilon_{el})} \quad \left[\frac{\mathrm{kg_{CO_2}}}{\mathrm{MWh}}\right] \end{aligned} \tag{4.13}$$

where the CO_2 emissions expressed in kilograms from a unit of chemical energy burned sequentially from coal and gas are respectively: $\rho_{CO_2}^{coal} \cong 95\ \mathrm{kg_{CO_2}/GJ}$, $\rho_{CO_2}^{gas} \cong 55\ \mathrm{kg_{CO_2}/GJ}$.

For a block in which only one fuel is burned, the emission factor is given by:

$$EF_{CO_2} = \frac{E_{ch}\rho_{CO_2}}{E_{el}} = \frac{\rho_{CO_2}}{\eta_{el}} \tag{4.14}$$

where:

η_{el} the net efficiency of electricity generating in a block,

and for example for a coal block for supercritical parameters with a net efficiency $\eta_{el} = 45.6\%$ (as already indicated above, this efficiency is achievable in blocks for supercritical parameters with values of at least 28 MPa, 600/620 °C; unit investment outlays are then $i \cong 6.5$ mln PLN/MW) the value of this factor is $EF_{CO_2} = 750\,\mathrm{kg_{CO_2}/MWh}$, and for a hierarchical gas-steam unit with net efficiency $\eta_{el} = 55\%$ assumes only the $EF_{CO_2} = 360\,\mathrm{kg_{CO_2}/MWh}$ value. In the dual-fuel gas-steam combined cycle, therefore, the EF_{CO_2} factor will be more closely related to the CO_2 emission factor for the hierarchical system, the more it will burn the gas in relation to the coal burned in the steam boiler, so the higher the q_{par} value.

The value of the EF_{CO_2} factor for the dual-fuel gas-steam combined cycle as a function is shown in Fig. 4.6. The following input data values were used for the calculations: $\eta_{GT} = 0.35$, $\eta_{HRSG} = 0.85$, $\eta_B = 0.9$, $\eta_{col} = 0.98$, $\eta_{me} = 0.98$, $\varepsilon_{el} = 0.04$. The value of the energy efficiency of the *Clausius-Rankine* cycle of the steam turbine was assumed once for the subcritical parameters of fresh steam equal to $\eta_{ST} = 0.45$, and two, for supercritical parameters equal to $\eta_{ST} = 0.55$.

4.1.1 Discussion and Analysis of the Results of Exemplary Calculations

Figures 4.1, 4.1a, 4.2, 4.2a and 4.3 presents sample results of calculations of the average specific cost of electricity production $k_{el,av}$ formula (4.8) in a steam-gas dual-fuel power unit. They are presented for two variants of investment outlays per block, once for $i \cong 4.6$ mln PLN/MW, that is when the steam part works with supercritical parameters, and two when $i \cong 3.4$ mln PLN/MW, so for subcritical parameters. In the first case, the efficiency of the *Clausius-Rankine* cycle was equal to $\eta_{ST} = 0.55$, in the second $\eta_{ST} = 0.45$. Moreover, the results are presented for the economic parameters that have the greatest impact on this cost. In Fig. 4.4, the boundary of gas e^{bo}_{gas} with the boundary price of coal e^{bo}_{coal} is also presented as a parameter, i.e. prices for which the cost $k_{el,av}$ assumes constant values, independent of value q_{par}. These lines are therefore horizontal in function q_{par}, as in Figs. 4.1, 4.1a, 4.2, 4.2a, 4.3 and 4.3a. Figure 4.6 shows the change in the value of the carbon dioxide emission factor EF_{CO_2} from the block. All calculation results presented in the paper correspond to zero values of all exponents a_{gas}, a_{coal}, a_{CO_2}, a_{CO}, a_{SO_2}, a_{NO_X}, a_{dust}, b_{CO_2} [1, 2]. Thus, prices of energy carriers and environmental fees are constant values in the entire T period of the block's operation, so they are total mean values in the T range. For the calculation it was also assumed that the ratio of chemical energy of fuel to its total use, for which it is not necessary to purchase CO_2 emission allowances is equal to zero, $u = 0$, as from year 2020 there will be no longer free allocations and power plants will have to pay for each emitted tonne of carbon dioxide.

As shown by curves on Figs. 4.1, 4.1a, 4.2, 4.2a, 4.3 and 4.3a, the block for the subcritical parameters of the fresh steam is more economically profitable. Only for the high price e_{CO_2} of purchasing CO_2 allowances, supercritical parameters are more profitable (Fig. 4.3, curve 1). This price has fluctuated within the last few years, ranging from $e_{CO_2} = 20$ PLN/Mg_{CO2} to $e_{CO_2} = 30$ PLN/Mg_{CO2}. Subcritical parameters of fresh steam are also more economically profitable if coal and gas prices are taken into account.

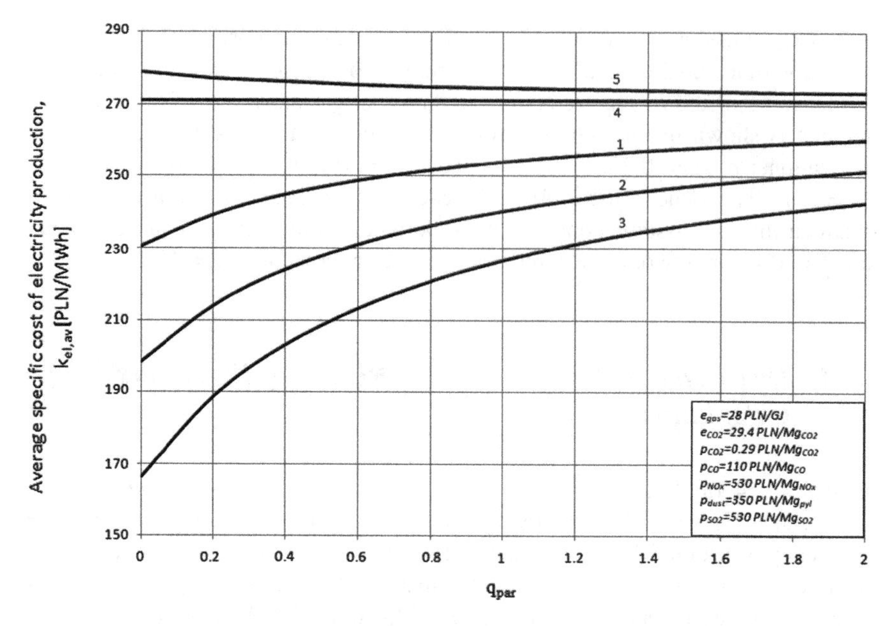

Fig. 4.2 Average specific cost of electricity production q_{par} in function with the price of coal as a parameter, where: 1 applies to $e_{coal} = 14$ PLN/GJ, i = 4.6 mln PLN/MW, $\eta_{ST} = 0.55$; 2 applies to $e_{coal} = 10$ PLN/GJ, i = 4.6 mln PLN/MW, $\eta_{ST} = 0.55$; 3 applies to $e_{coal} = 6$ PLN/GJ; i = 4.6 mln PLN/MW, $\eta_{ST} = 0.55$; 4 applies to $e^{bo}_{coal} = 19.05$ PLN/GJ, i = 4.6 mln PLN/MW, $\eta_{ST} = 0.55$; 5 applies to $e_{coal} = 20$ PLN/GJ, i = 4.6 mln PLN/MW, $\eta_{ST} = 0.55$

Currently, the price of coal in Poland is approx. $e_{coal} = 11.4$ PLN/GJ, gas price around $e_{gas} = 28$ PLN/GJ.

In Figs. 4.1 and 4.1a, the curves 2, 3 courses are decreasing because costs related to gas in Eq. (4.10) are lower than costs associated with coal. The values of these curves are calculated for gas prices $e_{gas} = 16$ PLN/GJ and $e_{gas} = 8$ PLN/GJ, lower than its boundary prices equal to: $e^{bo}_{gas} = 20.01$ PLN/GJ for $\eta_{ST} = 0.45$ and $e^{bo}_{gas} = 17.67$ PLN/GJ for $\eta_{ST} = 0.55$. The gas boundary prices correspond to the current price of coal equal to $e_{coal} = 11.4$ PLN/GJ and current purchase price of CO_2 emission allowances equal to $e_{CO_2} = 29.4$ PLN/Mg. Curves 1 in Figs. 4.1 and 4.1a. however, have a growing character, because the price of gas $e_{gas} = 32$ PLN/GJ is higher for them than the boundary prices, and therefore the costs associated with gas in Eq. (4.10) exceed the costs associated with coal.

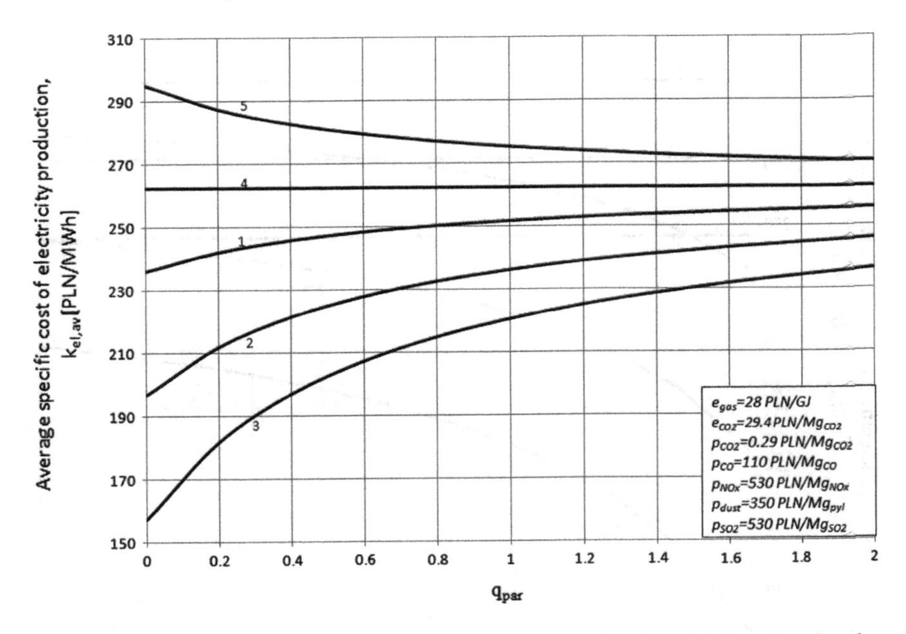

Fig. 4.2a Average specific cost of electricity production q_{par} in function with the price of coal as a parameter, where: 1 applies to e_{coal} = 14 PLN/GJ; i = 3.4 mln PLN/MW, η_{ST} = 0.45; 2 applies to e_{coal} = 10 PLN/GJ; i = 3.4 mln PLN/MW, η_{ST} = 0.45; 3 applies to e_{coal} = 6 PLN/GJ, i = 3.4 mln PLN/MW, η_{ST} = 0.45; 4 applies to e^{bo}_{coal} = 16.68 PLN/GJ; i = 3.4 mln PLN/MW, η_{ST} = 0.45; 5 applies to e_{coal} = 20 PLN/GJ; i = 3.4 mln PLN/MW, η_{ST} = 0.45

As can be seen from the course of curves 2, 3 in Figs. 4.1 and 4.1a. the dual-fuel gas-steam combined cycle is more economically viable than a single coal block only if the gas price is lower than its boundary prices set for the current coal price. The specific cost of electricity production $k_{el,av}$ is then the smallest for the maximum power of the gas turbine set. For the current price of coal equal to the e_{coal} = 11.4 PLN/GJ, price of gas must be less than the prices: e^{bo}_{gas} = 20.01 PLN/GJ for η_{ST} = 0.45 and e^{gr}_{gas} = 17.67 PLN/GJ for η_{ST} = 0.55. The current price of gas in Poland is e_{gas} = 28 PLN/GJ so dual fuel gas-steam combined cycle in a parallel system are therefore unprofitable.

The curves 1, 2, 3 in Figs. 4.2 and 4.2a. have an increasing course, because the costs associated with gas in Eq. (4.10) outweigh the costs associated with coal. The values of these curves are calculated for coal prices e_{coal} lower than its boundary prices (curves 4 in Figs. 4.2 and 4.2a). The boundary prices of coal correspond to the current price of gas equal to e_{gas} = 28 PLN/GJ and current price for the purchase of CO_2 emission allowances e_{CO_2} = 29.4 PLN/Mg.

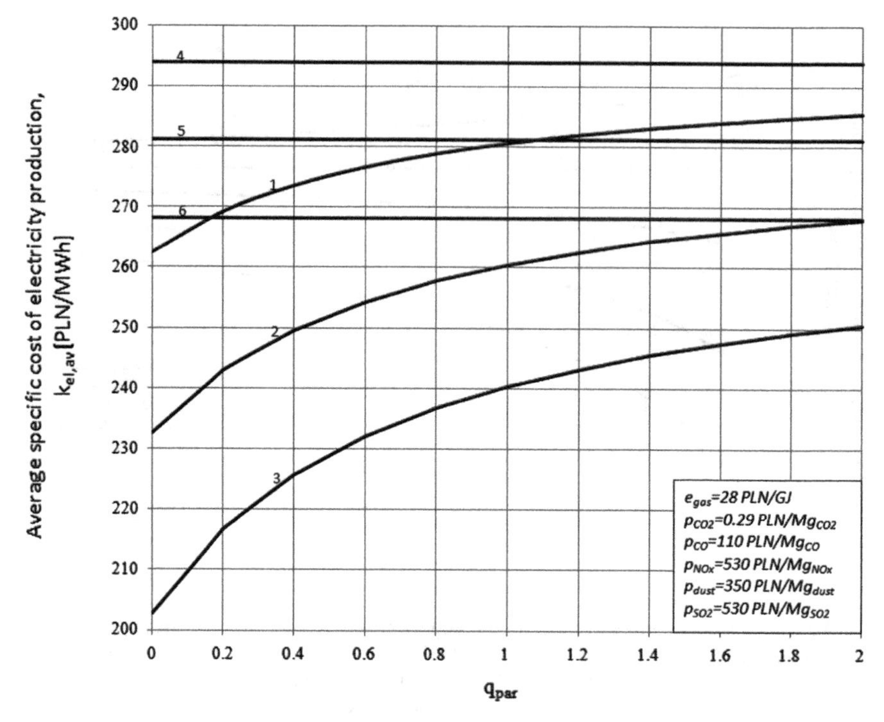

Fig. 4.3 Average specific cost of electricity production q_{par} in function with the purchase price of carbon dioxide emission allowances as a parameter, where: 1 applies to $e_{co_2} = 100$ PLN/Mg, i = 4.6 mln PLN/MW, $\eta_{ST} = 0.55$, $e_{węg} = 11.4$ PLN/GJ; 2 applies to $e_{co_2} = 60$ PLN/Mg, i = 4.6 mln PLN/MW, $\eta_{ST} = 0.55$, $e_{coal} = 11.4$ PLN/GJ; 3 applies to $e_{co_2} = 20$ PLN/Mg, i = 4.6 mln/MW, $\eta_{ST} = 0.55$, $e_{coal} = 11.4$ PLN/GJ; 4 applies to $e_{co_2} = 100$ PLN/Mg, i = 4.6 mln PLN/MW, $\eta_{ST} = 0.55$, $e^{bo}_{coal} = 15.29$ PLN/GJ; 5 applies to $e_{co_2} = 60$ PLN/Mg, i = 4.6 mln PLN/MW, $\eta_{ST} = 0.55$, $e^{bo}_{coal} = 17.42$ PLN/GJ; 6 applies to $e_{co_2} = 20$ PLN/Mg, i = 4.6 mln PLN/MW, $\eta_{ST} = 0.55$, $e^{bo}_{coal} = 19.55$ PLN/GJ

Curves 5 on the other hand, have a decreasing character because they were made for the price of coal $e_{coal} = 20$ PLN/GJ higher than the boundary ones, and therefore the relation $dk_{el,av} / dq_{par} < 0$ is fulfilled. The cost of electricity production $k_{el,av}$ is also the highest due to the highest price of coal and high gas price.

Figures 4.3 and 4.3a. shows the impact of the purchase price e_{CO_2} for the emission allowances of one tonne of carbon dioxide on the specific cost of electricity production $k_{el,av}$. The higher the price e_{CO_2}, the higher the boundary value e^{bo}_{coal}, so the higher the price of gas may occur for which the construction of dual-fuel gas-steam power blocks will be economically viable. This does not mean, however, that increasing the e_{CO_2} price is beneficial, it is quite the opposite. Increasing the e_{CO_2} price causes an increase in the $k_{el,av}$ cost.

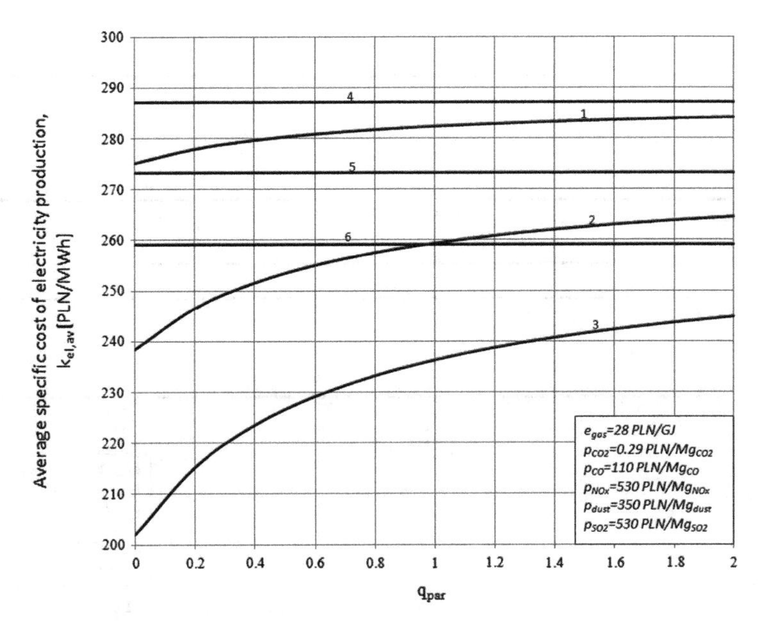

Fig. 4.3a Average specific cost of electricity production q_{par} in function with the purchase price of carbon dioxide emission allowances as a parameter, where: 1 applies to $e_{CO_2} = 100$ PLN/Mg, i = 3.4 mln PLN/MW, $\eta_{TP} = 0.45$, $e_{węg} = 11.4$ PLN/GJ; 2 applies to $e_{CO_2} = 60$ PLN/Mg, i = 3.4 mln/MW, $\eta_{ST} = 0.45$, $e_{coal} = 11.4$ PLN/GJ; 3 applies to $e_{CO_2} = 20$ PLN/Mg, i = 3.4 mln PLN/MW, $\eta_{ST} = 0.45$, $e_{coal} = 11.4$ PLN/GJ; 4 applies to $e_{CO_2} = 100$ PLN/Mg, i = 3.4 mln PLN/MW, $\eta_{ST} = 0.45$, $e^{bo}_{coal} = 12.62$ PLN/GJ; 5 applies to $e_{CO_2} = 60$ PLN/Mg, i = 3.4 mln PLN/MW, $\eta_{ST} = 0.45$, $e^{bo}_{coal} = 14.92$ PLN/GJ; 6 applies to $e_{CO_2} = 20$ PLN/Mg, i = 3.4 mln PLN/MW, $\eta_{ST} = 0.45$, $e^{gr}_{coal} = 17.22$ PLN/GJ

The curves 1, 2, 3 in Figs. 4.3 and 4.3a. are of increasing character for the same reason as the curves in Figs. 4.1, 4.1a, 4.2 and 4.2a. The values of these curves were calculated for the price of coal $e_{coal} = 11.4$ PLN/GJ lower than its boundary prices (curves 4, 5, 6). All boundary prices of coal correspond to the price of gas equal to $e_{gas} = 28$ PLN/GJ.

Figures 4.4 and 4.4a. shows the gas e^{bo}_{gas} and coal e^{bo}_{coal} boundary prices calculated using formula (4.10), i.e. prices for which the cost $k_{el,av}$ assumes a constant value independent of q_{par}. The higher the boundary coal price, the higher the boundary gas price, and vice versa. As it results from the necessary condition (4.11), the higher the price e^{bo}_{gas}, the higher may be the price e_{gas} for which the construction of dual-fuel gas-steam power blocks is profitable.

For example, the boundary price of coal $e^{bo}_{coal} = 10$ PLN/GJ corresponds to the boundary gas price equal to $e^{bo}_{gas} = 17.9$ PLN/GJ for $\eta_{ST} = 0.45$ and i = 3.4 mln PLN/MW, i.e. for a block with subcritical parameters of a fresh steam . For

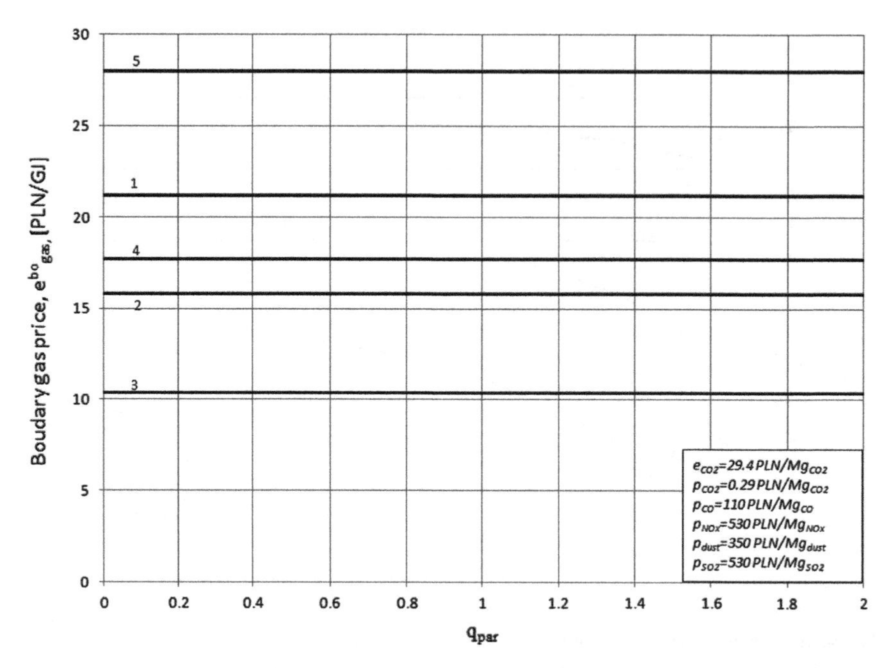

Fig. 4.4 The gas boundary price e^{bo}_{gas} as a function of q_{par} with the boundary price of coal e^{bo}_{coal} as a parameter, where: 1 applies to $e^{bo}_{coal} = 14$ PLN/GJ, i = 4.6 mln PLN/MW, $\eta_{ST} = 0.55$; 2 applies to $e^{bo}_{coal} = 10$ PLN/GJ, i = 4.6 mln PLN/MW, $\eta_{ST} = 0.55$; 3 applies to $e^{bo}_{coal} = 6$ PLN/GJ; i = 4.6 mln PLN/MW, $\eta_{ST} = 0.55$; 4 applies to $e^{bo}_{coal} = 11.4$ PLN/GJ, i = 4.6 mln PLN/MW, $\eta_{ST} = 0.55$; 5 applies to $e^{bo}_{coal} = 19.05$ PLN/GJ, i = 4.6 mln PLN/MW, $\eta_{ST} = 0.55$

$e^{bo}_{coal} = 10$ PLN/GJ and $\eta_{ST} = 0.55$ and $i = 4.6$ mln PLN/MW the boundary price of gas is $e^{bo}_{gas} = 15.8$ PLN/GJ et cetera. The above prices were calculated for the current purchase price of CO_2 emission allowances equal to $e_{CO_2} = 29.4$ PLN/Mg.

Figures 4.5 and 4.5a. presents a wide range of the gas price boundary value e^{bo}_{gas} as a function of the boundary price of coal e^{bo}_{coal} with the purchase price of carbon dioxide emission allowances e_{CO_2} and the energy efficiency of the steam turbine cycle η_{ST} as a parameter.

Figure 4.6 presents curves showing the change in the value of the carbon dioxide emission factor EF_{CO_2} from a dual-fuel gas-steam power unit in a parallel system per megawatt hour of electric energy produced in it formula (4.13) and gross efficiency $\eta_{el}^{GS,df}$ curves of electricity generation formula (4.4). It is obvious that the higher the energy efficiency η_{ST} of a steam part of the dual-fuel steam-gas power block, the lower the value of the indicator EF_{CO_2}, and the higher the efficiency $\eta_{el}^{GS,df}$.

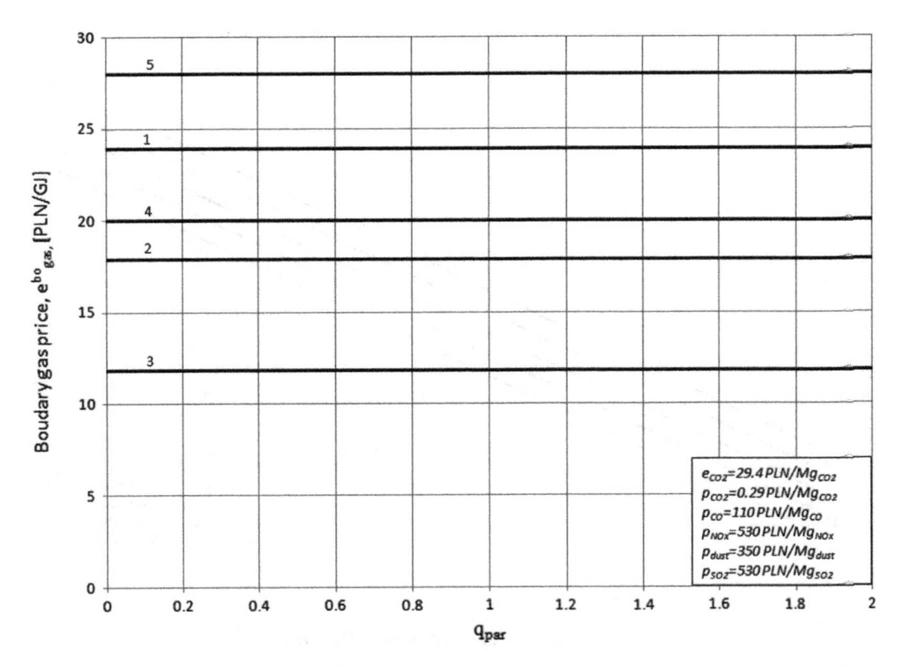

Fig. 4.4a The gas boundary price e^{bo}_{gas} as a function of q_{par} with the boundary price of coal e^{bo}_{coal} as a parameter, where: 1 applies to e^{bo}_{coal} = 14 PLN/GJ; i = 3.4 mln PLN/MW, η_{ST} = 0.45; 2 applies to e^{bo}_{coal} = 10 PLN/GJ; i = 3.4 mln PLN/MW, η_{ST} = 0.45; 3 applies to e^{bo}_{coal} = 6 PLN/GJ, i = 3.4 mln PLN/MW, η_{ST} = 0.45; 4 applies to e^{bo}_{coal} = 11.4 PLN/GJ; i = 3.4 mln PLN/MW, η_{ST} = 0.45; 5 applies to e^{bo}_{coal} = 16.68 PLN/GJ; i = 3.4 mln PLN/MW, η_{ST} = 0.45

4.1.2 Summary

Using the methodologies and mathematical models presented in the monograph for dual-fuel gas-steam combined cycle technology in a parallel system, multivariate calculations are presented in Figs. 4.1, 4.1a, 4.2, 4.2a, 4.3, 4.3a, 4.4, 4.4a, 4.5, 4.5a and 4.6. They were presented for a wide range of price changes of energy carriers and environmental fees, and therefore also for current prices and charges, as well as for various investment outlays. All the figures were obtained using, what is important, innovative methodology and mathematical models of total, discounted for the current moment NPV profit achieved from the operation of the dual-fuel gas-steam power unit with a continuous time formula (3.1). The decisive values of the power of the gas turbine set $N^{GT}_{el\,\max}$ are gas, coal, environmental costs and capital expenditures (capital expenditures determine value of η_{ST}). The lower the prices and expenditures, the lower is of course the specific cost of electricity generation $k_{el,av}$ formula (4.8).

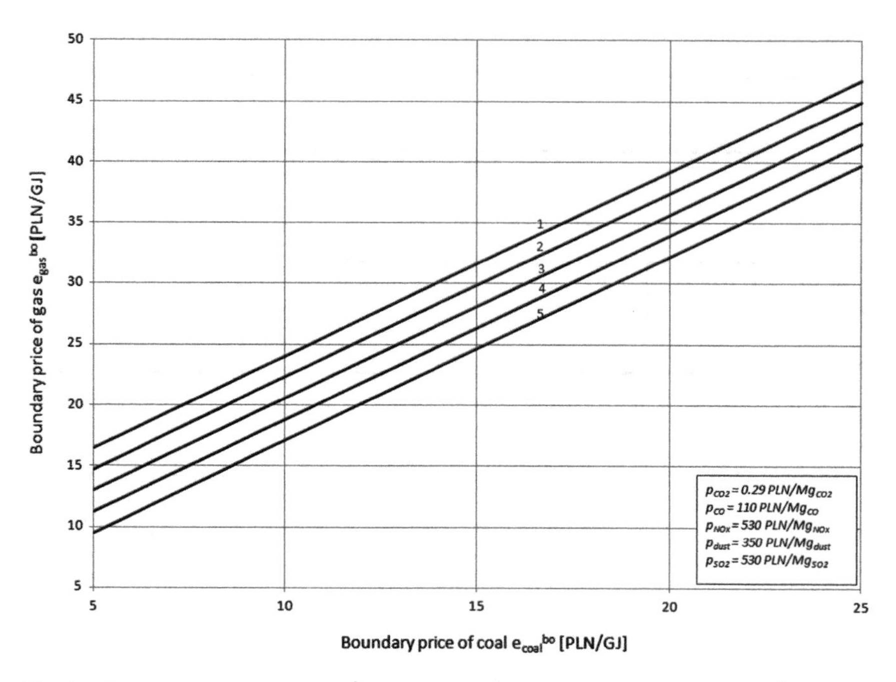

Fig. 4.5 The boundary price of gas e^{bo}_{gas} as a function of the boundary price of coal e^{bo}_{coal} and the price of CO_2 emissions e_{CO_2} as a parameter, where: 1 applies to $e_{CO_2} = 100$ PLN/Mg, i = 3.4 mln PLN/MW, $\eta_{TP} = 0.45$; 2 applies to $e_{CO_2} = 80$ PLN/Mg, i = 3.4 mln PLN/MW, $\eta_{TP} = 0.45$; 3 applies to $e_{CO_2} = 60$ PLN/Mg, i = 3.4 mln PLN/MW, $\eta_{TP} = 0.45$; 4 applies to $e_{CO_2} = 40$ PLN/Mg, i = 3.4 mln PLN/MW, $\eta_{TP} = 0.45$; 5 applies to $e_{CO_2} = 20$ PLN/Mg, i = 3.4 mln PLN/MW, $\eta_{TP} = 0.45$

The necessary condition for the economic viability of dual-fuel gas-steam power blocks, and hence their higher profitability than coal power plants, is the fulfillment of a relation $e_{gas} \leq e^{bo}_{gas}$, therefore a relation that guarantees lower costs associated with gas than costs associated with coal formula (4.9). The lower the gas price than its boundary price (determined for the current coal price), the lower is the specific cost of electricity production $k_{el,av}$ and the highest power of the gas turbine set is the most beneficial—curves 2, 3 in Figs. 4.1 and 4.1a. The higher the purchase price e_{CO_2} of CO_2 emission allowances, the higher the boundary value e^{bo}_{coal} (Figs. 4.3 and 4.3a), and thus the gas price formula (4.11), for which the construction of dual-fuel gas-steam power units will be profitable, may be higher. However, it does not mean, as already mentioned above, that increasing the e_{CO_2} price is beneficial. It is absolutely the opposite, because increasing the e_{CO_2} price, which the European Union aims at, increases the cost of electricity production $k_{el,av}$, and thus empties the pockets of electricity users to a greater extent.

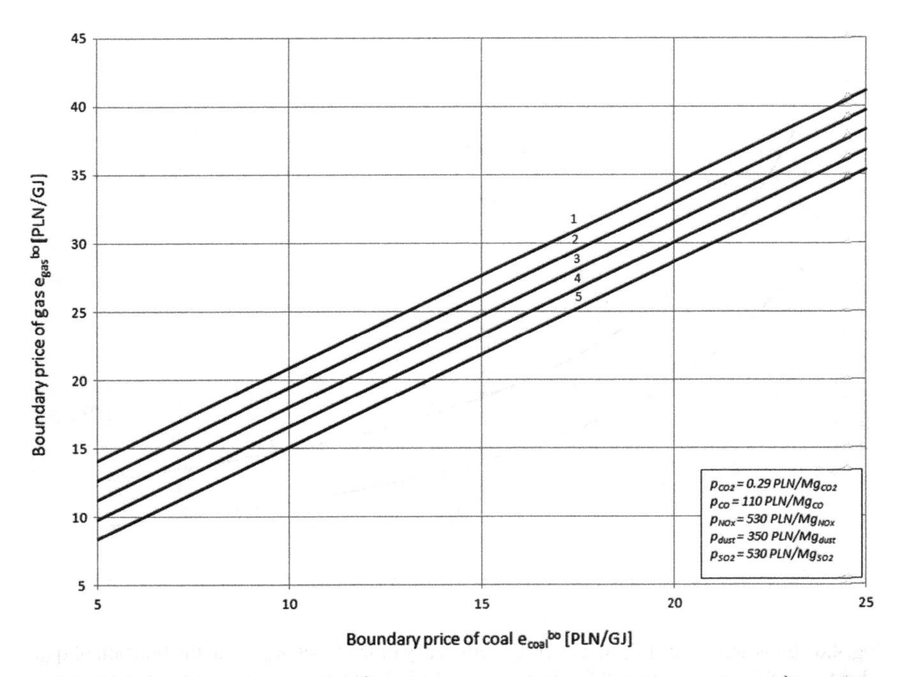

Fig. 4.5a The boundary price of gas e_{gas}^{bo} as a function of the boundary price of coal e_{coal}^{bo} and the price of CO_2 emissions e_{CO_2} as a parameter, where: 1 applies to e_{CO_2} = 100 PLN/Mg, i = 4.6 mln PLN/MW, η_{TP} = 0.55; 2 applies to e_{CO_2} = 80 PLN/Mg, i = 4.6 mln PLN/MW, η_{TP} = 0.55; 3 applies to e_{CO_2} = 60 PLN/Mg, i = 4.6 mln PLN/MW, η_{TP} = 0.55; 4 applies to e_{CO_2} = 40 PLN/Mg, i = 4.6 mln PLN/MW, η_{TP} = 0.55; 5 applies to e_{CO_2} = 20 PLN/Mg, i = 4.6 mln PLN/MW, η_{TP} = 0.55

4.2 Selection of the Optimum Power of the Gas Turbine Set for the Newly Built Dual-Fuel Gas-Steam Combined Heat and Power Plant in a Parallel System with a Bleed-Backpressure Turbine

In a dual-fuel gas-steam combined heat and power plant in a parallel system with a bleed-backpressure steam turbine (Fig. 3.6), the desired form of the average unit heat production cost, i.e. in the function of the searching value q_{par}, is given by:

$$
\begin{aligned}
k_{h,av} = \frac{r}{1-\mathrm{e}^{-rT}} \Bigg\{ & \frac{q_{par}}{[q_{par}(1-\eta_{GT})\eta_{HRSG}+\eta_B]\eta_{SH}(1-\eta_{ST}^{p})\eta_{HE}} \Bigg\{ (1+x_{sw,m,was}) e_{gas}^{t=0} \frac{1}{a_{gas}-r}[\mathrm{e}^{(a_{gas}-r)T}-1] \\
& + \rho_{CO_2}^{gas} p_{CO_2}^{t=0} \frac{1}{a_{CO_2}-r}[\mathrm{e}^{(a_{CO_2}-r)T}-1] + \rho_{CO}^{gas} p_{CO}^{t=0} \frac{1}{a_{CO}-r}[\mathrm{e}^{(a_{CO}-r)T}-1] \\
& + \rho_{NO_X}^{gas} p_{NO_X}^{t=0} \frac{1}{a_{NO_X}-r}[\mathrm{e}^{(a_{NO_X}-r)T}-1] + \rho_{SO_2}^{gas} p_{SO_2}^{t=0} \frac{1}{a_{SO_2}-r}[\mathrm{e}^{(a_{SO_2}-r)T}-1] \\
& + \rho_{dust}^{gas} p_{dust}^{t=0} \frac{1}{a_{dust}-r}[\mathrm{e}^{(a_{dust}-r)T}-1] + (1-u)\rho_{CO_2}^{gas} e_{CO_2}^{t=0} \frac{1}{b_{CO_2}-r}[\mathrm{e}^{(b_{CO_2}-r)T}-1] \Bigg\} \\
& + \frac{1}{[q_{par}(1-\eta_{GT})\eta_{HRSG}+\eta_B]\eta_{SH}(1-\eta_{ST}^{p})\eta_{HE}} \Bigg\{ (1+x_{sw,m,was}) e_{coal}^{t=0} \frac{1}{a_{coal}-r}[\mathrm{e}^{(a_{coal}-r)T}-1]
\end{aligned}
$$

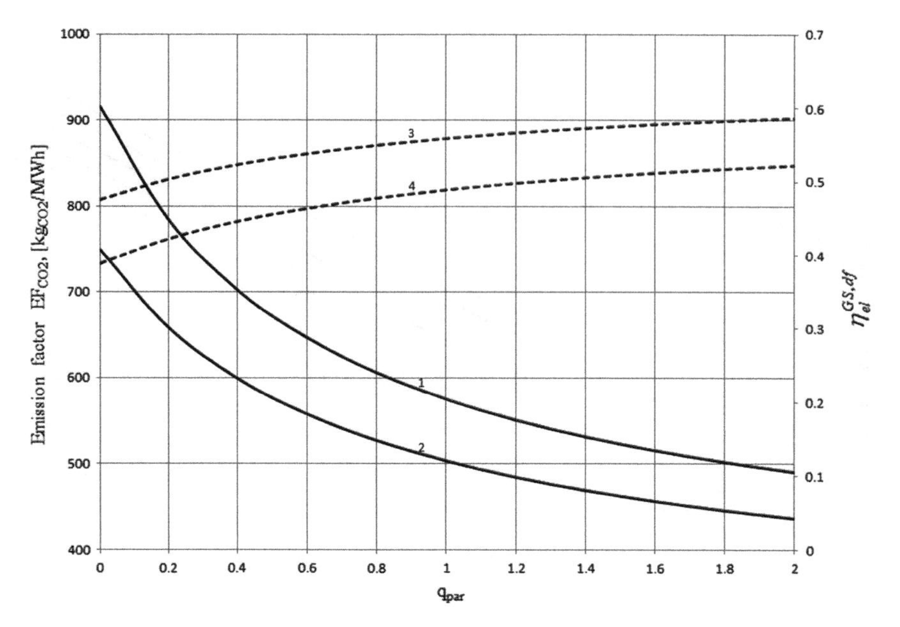

Fig. 4.6 Emission factor EF_{CO_2} and energy efficiency of the block $\eta_{el}^{GS,df}$ in the function of q_{par}, where: 1 applies to the emission factor for $\eta_{ST} = 0.45$; 2 applies to the emission factor for $\eta_{ST} = 0.55$; 3 concerns efficiency for $\eta_{ST} = 0.55$; 4 concerns efficiency for $\eta_{ST} = 0.45$

$$+ \rho^{coal}_{CO_2} p^{t=0}_{CO_2} \frac{1}{a_{CO_2}-r}[e^{(a_{CO_2}-r)T}-1] + \rho^{coal}_{CO} p^{t=0}_{CO} \frac{1}{a_{CO}-r}[e^{(a_{CO}-r)T}-1]$$

$$+ \rho^{coal}_{NO_X} p^{t=0}_{NO_X}\frac{1}{a_{NO_X}-r}[e^{(a_{NO_X}-r)T}-1] + \rho^{coal}_{SO_2}p^{t=0}_{SO_2}\frac{1}{a_{SO_2}-r}[e^{(a_{SO_2}-r)T}-1]$$

$$+ \rho^{coal}_{dust}p^{t=0}_{dust}\frac{1}{a_{dust}-r}[e^{(a_{dust}-r)T}-1] + (1-u)\rho^{coal}_{CO_2}e^{t=0}_{CO_2}\frac{1}{b_{CO_2}-r}[e^{(b_{CO_2}-r)T}-1]\Big\}$$

$$+\frac{i}{\tau_s}(1-e^{-rT})(1+x_{sal,t,ins})\frac{\delta_{serv}}{r}+\frac{zi}{\tau_s}\left(\frac{1-e^{-rT}}{T}+1\right)$$

$$-\frac{[q_{par}(1-\eta_{GT})\eta_{HRSG}+\eta_B]\eta_{SH}\eta^p_{ST}\eta_{me}+q_{par}\eta_{GT}}{[q_{par}(1-\eta_{GT})\eta_{HRSG}+\eta_B]\eta_{SH}(1-\eta^p_{ST})\eta_{HE}}(1-\varepsilon_{el})e^{t=0}_{el}\frac{1}{a_{el}-r}[e^{(a_{el}-r)T}-1]\Big\} \quad (4.15)$$

where:

- i specific (per unit of power) investment outlay per CHP plant, $i = J/\dot{Q}^{CHP}_{h\,max}$, (its value depends on the associated technology used for generation of heat and electricity),
- τ_s annual time of using thermal maximum power (peak) heat and power plant $\dot{Q}^{CHP}_{h\,max}$; $Q_A = \dot{Q}^{CHP}_{h\,max}\tau_s$

This formula was obtained from Eq. (3.5) after substituting the dependencies resulting from the energy balance presented in Fig. 3.6:

$$\frac{E_{ch,A}^{gas}}{Q_A} = \frac{q_{par}}{[q_{par}(1-\eta_{GT})\eta_{HRSG} + \eta_B]\eta_{SH}(1-\eta_{ST}^p)\eta_{HE}} \tag{4.16}$$

$$\frac{E_{ch,A}^{coal}}{Q_A} = \frac{1}{[q_{par}(1-\eta_{GT})\eta_{HRSG} + \eta_B]\eta_{SH}(1-\eta_{ST}^p)\eta_{HE}}, \tag{4.17}$$

$$\frac{E_{el,A}^{ST} + E_{el,A}^{GT}}{Q_A} = \frac{[q_{par}(1-\eta_{GT})\eta_{HRSG} + \eta_B]\eta_{SH}\eta_{ST}^p\eta_{me} + q_{par}\eta_{GT}}{[q_{par}(1-\eta_{GT})\eta_{HRSG} + \eta_B]\eta_{SH}(1-\eta_{ST}^p)\eta_{HE}}. \tag{4.18}$$

When calculating from formula (4.15) the derivative $dk_{h,av}/dq_{par}$ we obtained:

$$\begin{aligned}
\frac{dk_{h,av}}{dq_{par}} &= \frac{r}{1-e^{-rT}}\left\{\frac{\eta_B\eta_{SH}(1-\eta_{ST}^p)\eta_{HE}}{\{[q_{par}(1-\eta_{GT})\eta_{HRSG}+\eta_B]\eta_{SH}(1-\eta_{ST}^p)\eta_{HE}\}^2}\right.\\
&\times\left\{(1+x_{sw,m,was})e_{gas}^{t=0}\frac{1}{a_{gas}-r}[e^{(a_{gas}-r)T}-1]\right.\\
&+\rho_{CO_2}^{gas}p_{CO_2}^{t=0}\frac{1}{a_{CO_2}-r}[e^{(a_{CO_2}-r)T}-1]+\rho_{CO}^{gas}p_{CO}^{t=0}\frac{1}{a_{CO}-r}[e^{(a_{CO}-r)T}-1]\\
&+\rho_{NO_X}^{gas}p_{NO_X}^{t=0}\frac{1}{a_{NO_X}-r}[e^{(a_{NO_X}-r)T}-1]+\rho_{SO_2}^{gas}p_{SO_2}^{t=0}\frac{1}{a_{SO_2}-r}[e^{(a_{SO_2}-r)T}-1]\\
&\left.+\rho_{dust}^{gas}p_{dust}^{t=0}\frac{1}{a_{dust}-r}[e^{(a_{dust}-r)T}-1]+(1-u)\rho_{CO_2}^{gas}e_{CO_2}^{t=0}\frac{1}{b_{CO_2}-r}[e^{(b_{CO_2}-r)T}-1]\right\}\\
&-\frac{(1-\eta_{GT})\eta_{HRSG}\eta_{SH}(1-\eta_{ST}^p)\eta_{HE}}{\{[q_{par}(1-\eta_{GT})\eta_{HRSG}+\eta_B]\eta_{SH}(1-\eta_{ST}^p)\eta_{HE}\}^2}\\
&\times\left\{(1+x_{sw,m,was})e_{coal}^{t=0}\frac{1}{a_{coal}-r}[e^{(a_{coal}-r)T}-1]\right.\\
&+\rho_{CO_2}^{coal}p_{CO_2}^{t=0}\frac{1}{a_{CO_2}-r}[e^{(a_{CO_2}-r)T}-1]+\rho_{CO}^{coal}p_{CO}^{t=0}\frac{1}{a_{CO}-r}[e^{(a_{CO}-r)T}-1]\\
&+\rho_{NO_X}^{coal}p_{NO_X}^{t=0}\frac{1}{a_{NO_X}-r}[e^{(a_{NO_X}-r)T}-1]+\rho_{SO_2}^{coal}p_{SO_2}^{t=0}\frac{1}{a_{SO_2}-r}[e^{(a_{SO_2}-r)T}-1]\\
&\left.+\rho_{dust}^{coal}p_{dust}^{t=0}\frac{1}{a_{dust}-r}[e^{(a_{dust}-r)T}-1]+(1-u)\rho_{CO_2}^{coal}e_{CO_2}^{t=0}\frac{1}{b_{CO_2}-r}[e^{(b_{CO_2}-r)T}-1]\right\}\\
&+\left\{\frac{(1-\eta_{GT})\eta_{HRSG}\eta_{SH}(1-\eta_{ST}^p)\eta_{HE}\{[q_{par}(1-\eta_{GT})\eta_{HRSG}+\eta_B]\eta_{SH}\eta_{ST}^p\eta_{me}+q_{par}\eta_{GT}\}}{\{[q_{par}(1-\eta_{GT})\eta_{HRSG}+\eta_B]\eta_{SH}(1-\eta_{ST}^p)\eta_{HE}\}^2}\right.\\
&\left.-\frac{[(1-\eta_{GT})\eta_{HRSG}\eta_{SH}\eta_{ST}^p\eta_{me}+\eta_{GT}][q_{par}(1-\eta_{GT})\eta_{HRSG}+\eta_B]\eta_{SH}(1-\eta_{ST}^p)\eta_{HE}}{\{[q_{par}(1-\eta_{GT})\eta_{HRSG}+\eta_B]\eta_{SH}(1-\eta_{ST}^p)\eta_{HE}\}^2}\right\}\\
&\left.\times(1-\varepsilon_{el})e_{el}^{t=0}\frac{1}{a_{el}-r}[e^{(a_{el}-r)T}-1]\right\}
\end{aligned} \tag{4.19}$$

and then from the precondition for the existence of an extreme $dk_{h,av}/dq_{par} = 0$, the value of the extremist q_{par}^{ext} cost $k_{h,av}$ is determined:

$$\begin{aligned}
0 = \eta_B&\left\{(1+x_{sw,m,was})e_{gas}^{t=0}\frac{1}{a_{gas}-r}[e^{(a_{gas}-r)T}-1]\right.\\
&+\rho_{CO_2}^{gas}p_{CO_2}^{t=0}\frac{1}{a_{CO_2}-r}[e^{(a_{CO_2}-r)T}-1]+\rho_{CO}^{gas}p_{CO}^{t=0}\frac{1}{a_{CO}-r}[e^{(a_{CO}-r)T}-1]
\end{aligned}$$

$$
\begin{aligned}
&+\rho_{NO_X}^{gas} p_{NO_X}^{t=0}\frac{1}{a_{NO_X}-r}[e^{(a_{NO_X}-r)T}-1]+\rho_{SO_2}^{gas} p_{SO_2}^{t=0}\frac{1}{a_{SO_2}-r}[e^{(a_{SO_2}-r)T}-1]\\
&+\rho_{dust}^{gas} p_{dust}^{t=0}\frac{1}{a_{dust}-r}[e^{(a_{dust}-r)T}-1]+(1-u)\rho_{CO_2}^{gas} e_{CO_2}^{t=0}\frac{1}{b_{CO_2}-r}[e^{(b_{CO_2}-r)T}-1]\Big\}\\
&-(1-\eta_{GT})\eta_{HRSG}\Big\{(1+x_{sw,m,was})e_{coal}^{t=0}\frac{1}{a_{coal}-r}[e^{(a_{coal}-r)T}-1]\\
&+\rho_{CO_2}^{coal} p_{CO_2}^{t=0}\frac{1}{a_{CO_2}-r}[e^{(a_{CO_2}-r)T}-1]+\rho_{CO}^{coal} p_{CO}^{t=0}\frac{1}{a_{CO}-r}[e^{(a_{CO}-r)T}-1]\\
&+\rho_{NO_X}^{coal} p_{NO_X}^{t=0}\frac{1}{a_{NO_X}-r}[e^{(a_{NO_X}-r)T}-1]+\rho_{SO_2}^{coal} p_{SO_2}^{t=0}\frac{1}{a_{SO_2}-r}[e^{(a_{SO_2}-r)T}-1]\\
&+\rho_{dust}^{coal} p_{dust}^{t=0}\frac{1}{a_{dust}-r}[e^{(a_{dust}-r)T}-1]+(1-u)\rho_{CO_2}^{coal} e_{CO_2}^{t=0}\frac{1}{b_{CO_2}-r}[e^{(b_{CO_2}-r)T}-1]\Big\}\\
&-\eta_{GT}\eta_B(1-\varepsilon_{el})e_{el}^{t=0}\frac{1}{a_{el}-r}[e^{(a_{el}-r)T}-1]. \qquad (4.20)
\end{aligned}
$$

As can be seen from formula (4.20), the value q_{par}^{ext} extremizing cost $k_{c,s\,r}^{\prime}$ does not exist. This value in the formula (4.20) does not exist (this situation can change the dependence of unit outlays "i" on the value q_{par}). From the formula (4.20), however, it is possible to determine, for a given electricity price, the gas e_{gas}^{bo} and coal e_{coal}^{bo} boundary prices, i.e. prices, for which the cost $k_{h,av}$ assumes a constant value, and therefore independent of q_{par}, since then the derivative fulfills the condition $dk_{h,av}/dq_{par}=0$.

Thus, for a given price of coal e_{coal} and electricity e_{el} from the Eq. (4.20), a gas price boundary e_{gas}^{bo} is set, and vice versa, the boundary price of coal e_{coal}^{bo} is determined for a given price of gas e_{gas} and electricity e_{el}. Thus, the gas and coal boundary prices are closely related (Figs. 4.13 and 4.14) and substituted for the Eq. (4.19), they "zero" its value. On the other hand, if other values are inserted into Eq. (4.19), for example current coal and gas prices (current coal and gas prices in Poland are equal to $e_{coal}=11.4$ PLN/GJ, $e_{gas}=28$ PLN/GJ), different from boundary values, then the equation takes a positive or negative value.

If the gas costs in the Eq. (4.20) exceed the costs associated with coal, then $dk_{h,av}/dq_{par}>0$ and the cost $k_{h,av}$ increases with higher value of q_{par}, and vice versa, when the costs associated with gas are lower than the costs associated with coal, then the relation $dk_{h,av}/dq_{par}<0$ is met and the cost $k_{h,av}$ decreases with the increase in power gas turbine set. However, it should be noted that compliance $dk_{h,av}/dq_{par}<0$ is possible in two cases. Once, when the price of coal is higher than its boundary price (then the cost of heat production $k_{h,av}$ is highest, Figs. 4.8 and 4.11) and two, when the price of gas is lower than its boundary price (then the cost $k_{h,av}$ is the lowest, Figs. 4.7 and 4.10).

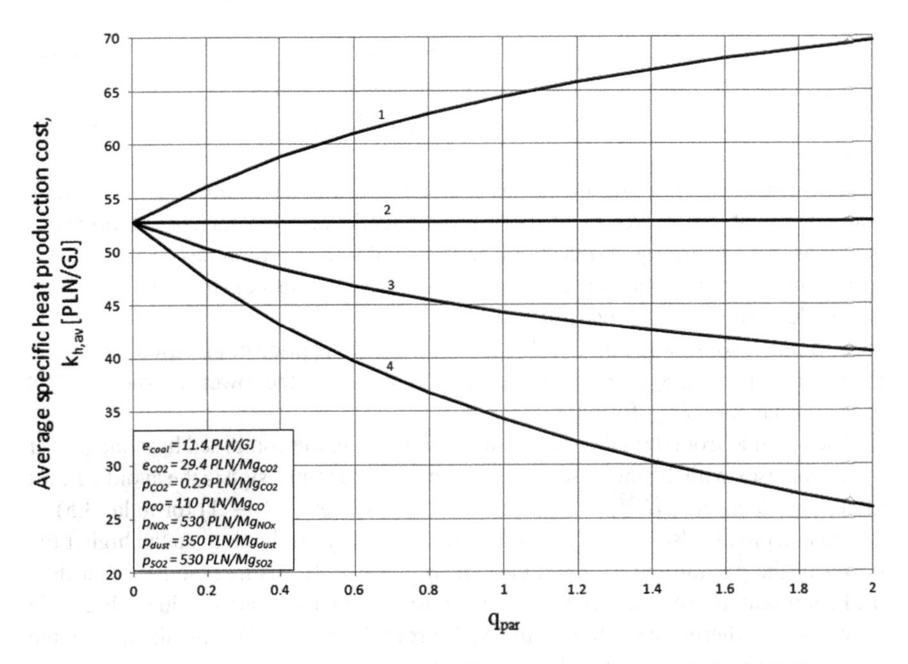

Fig. 4.7 Average specific cost of heat production as a function of q_{par} with the price of gas as a parameter, where: 1 applied to $e_{gas} = 32$ PLN/GJ, $e_{el} = 170$ PLN/MWh; 2 applied to $e_{gas}^{bo} = 22.71$ PLN/GJ; $e_{el} = 170$ PLN/MWh; 3 applied to $e_{gas} = 16$ PLN/GJ, $e_{el} = 170$ PLN/MWh; 4 applied to $e_{gas} = 8$ PLN/GJ, $e_{el} = 170$ PLN/MWh

Therefore, a necessary condition for a dual-fuel gas-steam CHP plant to be economically more profitable than a single coal-fired CHP plant (i.e. $q_{par} = 0$, Figs. 4.7 and 4.10), is not only the relation $dk_{h,av}/dq_{par} < 0$ that is met, but also that the gas price is lower than its boundary price determined from Eq. (4.20) for the current price of coal and electricity.

The necessary condition for the profitability of construction of dual-fuel CHP plants should therefore be finally recorded with the following relationship:

$$e_{gas} \leq e_{gas}^{bo}. \tag{4.21}$$

The higher the prices of electricity e_{el} and coal e_{coal}, the higher the boundary value e_{gas}^{bo}, and thus the price of gas may be higher, for which the construction of dual-fuel gas-steam CHP plant will be profitable (Figs. 4.13 and 4.14).

If the relationship $e_{gas} \leq e_{gas}^{bo}$ is fulfilled, then the most profitable is the maximum power of the gas turbine set, and therefore the greatest value $q_{par} = q_{par}^{\max}$ formula (4.12). Knowledge of the boundary price of coal e_{coal}^{bo} and gas e_{gas}^{bo} is therefore very important.

Knowing them, we are able to answer the question: is a dual-fuel gas-steam combined heat and power plant more economically justified for current prices of coal, gas and electricity and costs of emission of harmful products of the block's operation to the environment and investment outlays for its construction, or a more profitable is single coal-fired CHP plant?

It should also be noted that the higher the ratio of the price of electricity e_{el} to the prices of fuels e_{coal}, e_{gas} (especially expensive gas e_{gas}), the lower the specific cost of heat production $k_{h,av}$ formula (3.5).

The revenue from the sale of electricity produced in the combined heat and power plant increases with the increase of e_{el}, which with the minus sign is the avoiding cost of heat production: $-\left(E_{el,A}^{ST} + E_{el,A}^{GT}\right)(1 - \varepsilon_{el})e_{el}^{t=0}\frac{1}{a_{el}-r}[\mathrm{e}^{(a_{el}-r)T} - 1]$ formula (3.5).

This income is also bigger, the higher the electricity production, so the higher the power of the gas turbine, so the higher the q_{par} value. If this income is higher than the heat production costs, then the specific cost $k_{c,s'r}$ takes negative values - Fig. 4.7a curve 4—and therefore the higher the *NPV* profit formula (3.4) is obtained from the operation of combined heat and power plant.

In the calculations, it was assumed that specific outlays for a combined heat and power plant with extraction-condensing steam turbine of $i = 4.6$ million PLN/MW, and for a combined heat and power plant with extraction-backpressure turbine $i = 4$ million PLN/MW. At the same time, it was assumed that the efficiency of the *Clausius-Rankine* cycle of the extraction-condensing steam turbine is $\eta_{ST} = \eta_{CR}\eta_i = 0.45$, and the extraction-backpressure turbine equals $\eta_{ST} = \eta_{CR}\eta_i = 0.4$. The following input values were also adopted for calculations: b = 4 years, $\beta = 2$, $\eta_{GT} = 0.35$, $\eta_{HRSG} = 0.85$, $\eta_B = 0.9$, $\eta_{SH} = 0.98$, $\eta_{me} = 0.98$, $\eta_{HE} = 0.95$, $\varepsilon_{el} = 0.04$, $\delta_{serv} = 0.03$, $r = 7\%$, $\tau_s = 3000$ h/a, $T = 20$ years. All calculation results presented in the monograph correspond to zero values of all exponents a_{el}, a_{gas}, a_{coal}, $a_{\mathrm{CO_2}}$, a_{CO}, $a_{\mathrm{SO_2}}$, $a_{\mathrm{NO_X}}$, a_{dust}, $b_{\mathrm{CO_2}}$. Thus, prices of energy carriers and environmental fees are constant over the whole T operation period of a combined heat and power plant, thus they are integral mean values in the T period. For example, the integral mean value of the electricity price changing in time according to the formula $e_{el}(t) = e_{el}^{t=0}\mathrm{e}^{a_{el}t}$ is calculated from the formula:

$$e_{el}^{av} = \frac{1}{T}\int_{0}^{T} e_{el}^{t=0}\mathrm{e}^{a_{el}t}dt = \frac{e_{el}^{t=0}}{Ta_{el}}\left(\mathrm{e}^{a_{el}T} - 1\right) \tag{4.22}$$

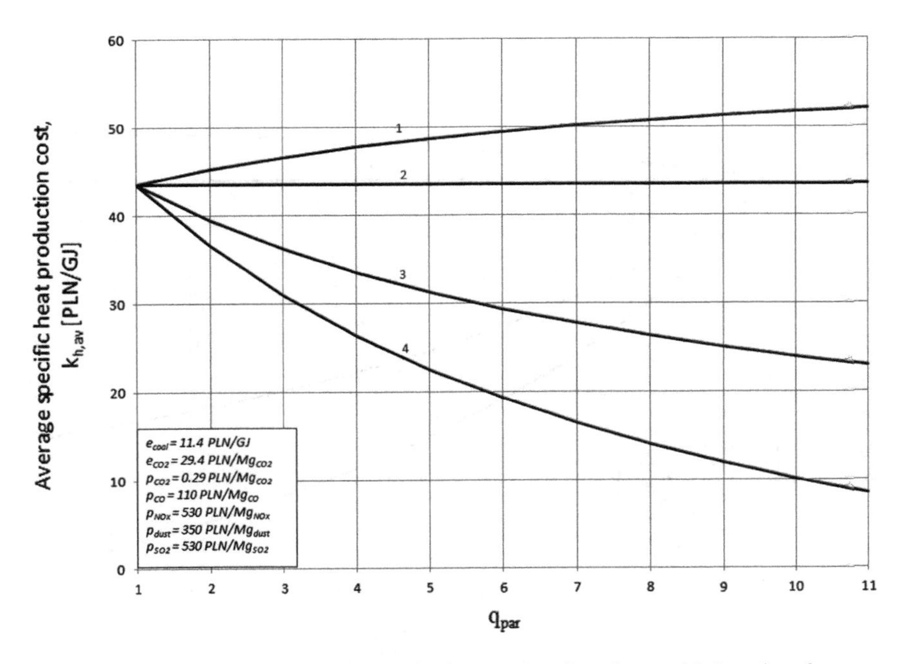

Fig. 4.7a Average specific cost of heat production as a function of q_{par} with the price of gas as a parameter, where: 1 $e_{gas} = 32$ PLN/GJ, $e_{el} = 220$ PLN/MWh; 2 $e_{gas}^{bo} = 27.29$ PLN/GJ; $e_{el} = 220$ PLN/MWh; 3 $e_{gas} = 16$ PLN/GJ, $e_{el} = 220$ PLN/MWh; 4 $e_{gas} = 8$ PLN/GJ, $e_{el} = 220$ PLN/MWh

The value of the chemical energy share of fuel in its total annual consumption was also used for calculations, for which CO_2 emission allowances are not required, $u = 0$, as from 2020 there will no longer be free allocations and installations will have to pay for each tonne of the carbon dioxide emitted.

4.2.1 Discussion and Analysis of Calculation Results

Figures 4.7, 4.7a, 4.7b, 4.8, 4.8a, 4.8b, 4.9, 4.9a, 4.9b, 4.10, 4.10a and 4.10b show the results of calculations of the specific cost of heat production $k_{h,av}$ for a combined heat and power plant with extraction-backpressure steam turbine. The character of the course of individual curves in these figures is identical to the courses of the corresponding curves in Figs. 4.10, 4.10a, 4.10b, 4.11, 4.11a, 4.11b, 4.12 and 4.12a prepared for a combined heat and power plant with extraction-condensing turbine. The only difference is lower cost values $k_{h,av}$ with the same energy prices and environmental fees.

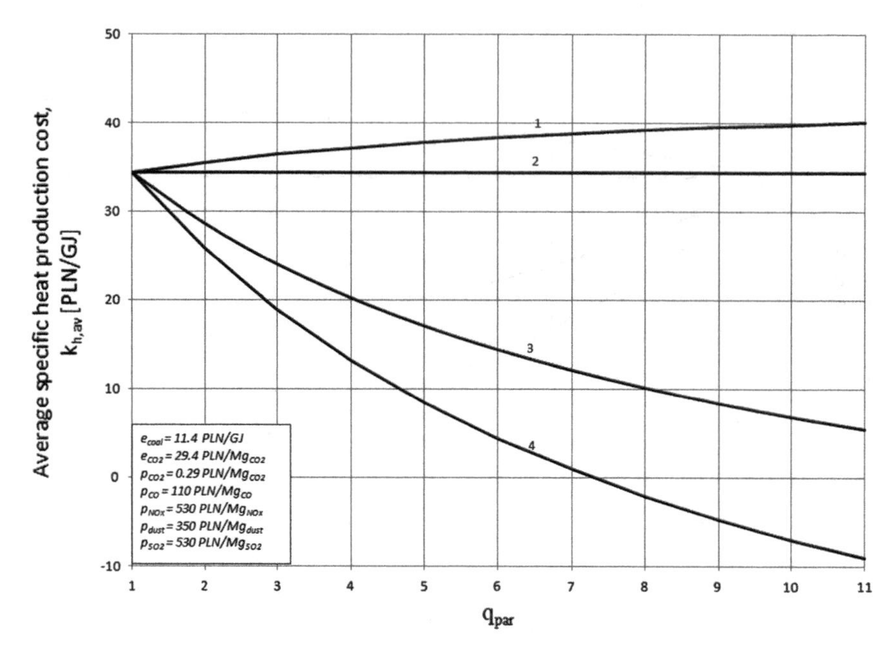

Fig. 4.7b Average specific cost of heat production as a function of q_{par} with the price of gas as a parameter, where: 1 $e_{gas} = 35$ PLN/GJ, $e_{el} = 270$ PLN/MWh, 2 $e_{gas}^{bo} = 31.86$ PLN/GJ; $e_{el} = 270$ PLN/MWh; 3 $e_{gas} = 16$ PLN/GJ, $e_{el} = 270$ PLN/MWh, 4 $e_{gas} = 8$ PLN/GJ, $e_{el} = 270$ PLN/MWh

This is because the total energy efficiency of the backpressure turbine cycle is greater than the efficiency of the condensing turbine, although it is smaller in it, with the same parameters of fresh steam, the range of use of steam enthalpy for the production of electricity. As mentioned above, the efficiency of electricity generation in the Clausius-Rankine cycle of the extraction-condensing steam turbine was assumed at the level of $\eta_{ST} = 0.45$, and for the cycle with the extraction-backpressure turbine equal to $\eta_{ST} = 0.4$.

The lower cost $k_{h,av}$, together with the higher total energy efficiency of the system with the extraction-backpressure turbine, is also the result of smaller specific investment outlays. In this system, as mentioned above, unnecessary are the under-turbine condenser and the cooling tower, which are necessary in the system with a condensing turbine.

The boundary prices of coal and gas for a given electricity price for combined heat and power plants with extraction backpressure steam turbine are also presented in Figs. 4.13 and 4.14. As already stated above, the gas and coal boundary prices do not depend on the type of steam turbine set installed in the CHP plant.

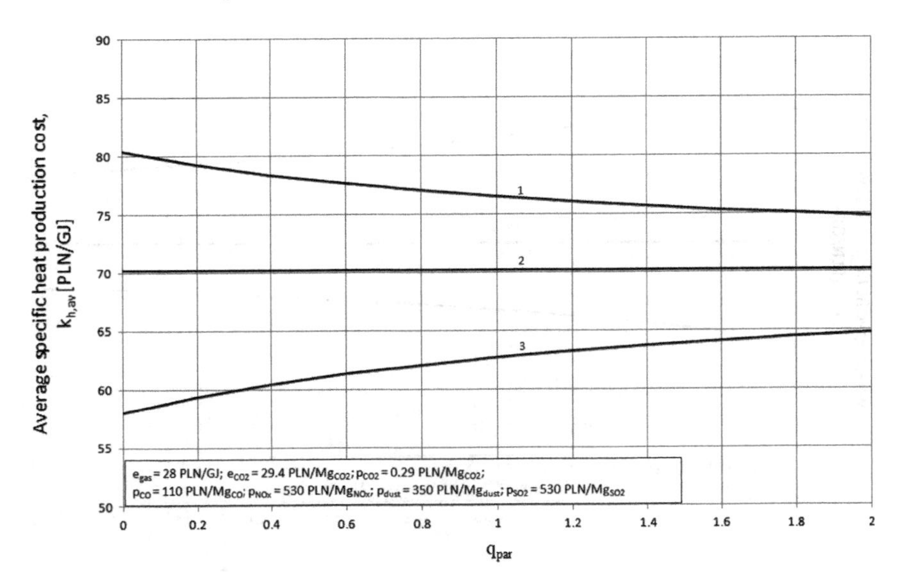

Fig. 4.8 Average specific cost of heat production as a function of q_{par} with the price of coal as a parameter, where: 1 applied to $e_{coal} = 25$ PLN/GJ, $e_{el} = 170$ PLN/MWh; 2 applied to $e^{bo}_{coal} = 20$ PLN/GJ, $e_{el} = 170$ PLN/MWh; 3 applied to $e_{coal} = 14$ PLN/GJ, $e_{el} = 170$ PLN/MWh

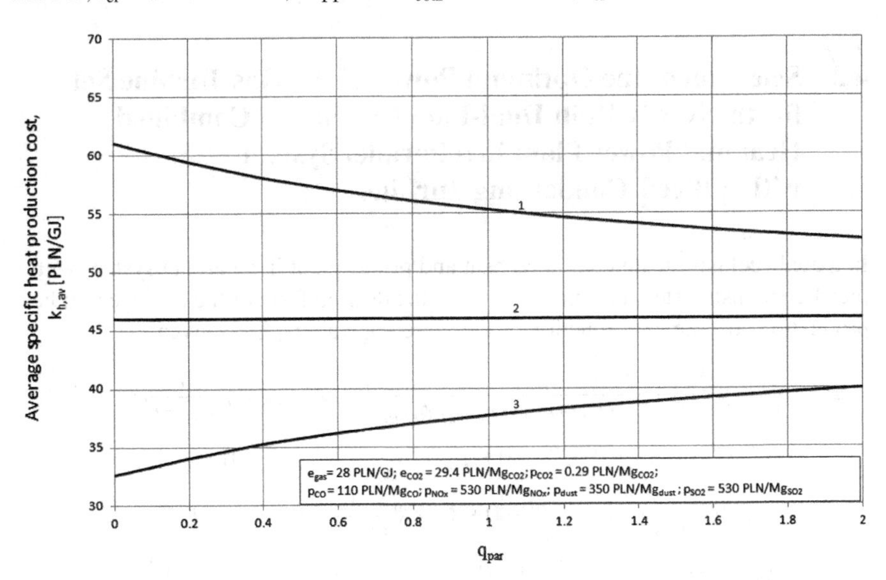

Fig. 4.8a Average specific cost of heat production as a function of q_{par} with the price of coal as a parameter, where: 1′ applied to $e_{coal} = 20$ PLN/GJ, $e_{el} = 220$ PLN/MWh; 2′ applied to $e^{bo}_{coal} = 12.55$ PLN/GJ; $e_{el} = 220$ PLN/MWh; 3′ applied to $e_{coal} = 6$ PLN/GJ, $e_{el} = 220$ PLN/MWh

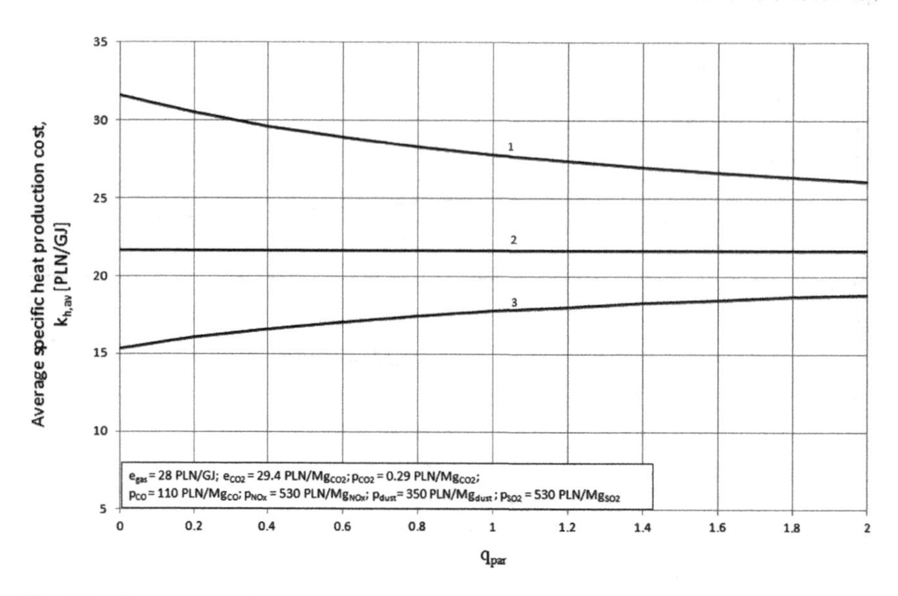

Fig. 4.8b Average specific cost of heat production as a function of q_{par} with the price of coal as a parameter, where: 1″ applied to $e_{coal} = 10$ PLN/GJ, $e_{el} = 270$ PLN/MWh; 2″ applied to $e_{coal}^{bo} = 5.11$ PLN/GJ, $e_{el} = 270$ PLN/MWh.; 3″ applied to $e_{coal} = 2$ PLN/GJ, $e_{el} = 270$ PLN/MWh

4.3 Selection of the Optimum Power of the Gas Turbine Set for the Newly Built Dual-Fuel Gas-Steam Combined Heat and Power Plant in a Parallel System with a Bleed-Condensing Turbine

In a dual-fuel gas-steam combined heat and power plant in a parallel system with a bleed-condensing steam turbine (Fig. 3.5), the desired form of the average unit heat production cost, i.e. as a function of the searching value q_{par}, is given by:

$$
\begin{aligned}
k_{h,av} = {} & \frac{r}{1-\mathrm{e}^{-rT}} \Bigg\{ \frac{q_{par}}{\beta[q_{par}(1-\eta_{GT})\eta_{HRSG}+\eta_B]\eta_{SH}\eta_{ST}^{k}\eta_{HE}} \Bigg\{ (1+x_{sw,m,was}) e_{gas}^{t=0} \frac{1}{a_{gas}-r}\left[\mathrm{e}^{(a_{gas}-r)T}-1\right] \\
& + \rho_{\mathrm{CO_2}}^{gas} p_{\mathrm{CO_2}}^{t=0} \frac{1}{a_{\mathrm{CO_2}}-r}\left[\mathrm{e}^{(a_{\mathrm{CO_2}}-r)T}-1\right] + \rho_{\mathrm{CO}}^{gas} p_{\mathrm{CO}}^{t=0} \frac{1}{a_{\mathrm{CO}}-r}\left[\mathrm{e}^{(a_{\mathrm{CO}}-r)T}-1\right] \\
& + \rho_{\mathrm{NO_X}}^{gas} p_{\mathrm{NO_X}}^{t=0} \frac{1}{a_{\mathrm{NO_X}}-r}\left[\mathrm{e}^{(a_{\mathrm{NO_X}}-r)T}-1\right] + \rho_{\mathrm{SO_2}}^{gas} p_{\mathrm{SO_2}}^{t=0} \frac{1}{a_{\mathrm{SO_2}}-r}\left[\mathrm{e}^{(a_{\mathrm{SO_2}}-r)T}-1\right] \\
& + \rho_{dust}^{gas} p_{dust}^{t=0} \frac{1}{a_{dust}-r}\left[\mathrm{e}^{(a_{dust}-r)T}-1\right] + (1-u)\rho_{\mathrm{CO_2}}^{gas} e_{\mathrm{CO_2}}^{t=0} \frac{1}{b_{\mathrm{CO_2}}-r}\left[\mathrm{e}^{(b_{\mathrm{CO_2}}-r)T}-1\right] \Bigg\} \\
& + \frac{1}{\beta[q_{par}(1-\eta_{GT})\eta_{HRSG}+\eta_B]\eta_{SH}\eta_{ST}^{k}\eta_{HE}} \Bigg\{ (1+x_{sw,m,was}) e_{coal}^{t=0} \frac{1}{a_{coal}-r}\left[\mathrm{e}^{(a_{coal}-r)T}-1\right] \\
& + \rho_{\mathrm{CO_2}}^{coal} p_{\mathrm{CO_2}}^{t=0} \frac{1}{a_{\mathrm{CO_2}}-r}\left[\mathrm{e}^{(a_{\mathrm{CO_2}}-r)T}-1\right] + \rho_{\mathrm{CO}}^{coal} p_{\mathrm{CO}}^{t=0} \frac{1}{a_{\mathrm{CO}}-r}\left[\mathrm{e}^{(a_{\mathrm{CO}}-r)T}-1\right] \\
& + \rho_{\mathrm{NO_X}}^{coal} p_{\mathrm{NO_X}}^{t=0} \frac{1}{a_{\mathrm{NO_X}}-r}\left[\mathrm{e}^{(a_{\mathrm{NO_X}}-r)T}-1\right] + \rho_{\mathrm{SO_2}}^{coal} p_{\mathrm{SO_2}}^{t=0} \frac{1}{a_{\mathrm{SO_2}}-r}\left[\mathrm{e}^{(a_{\mathrm{SO_2}}-r)T}-1\right]
\end{aligned}
$$

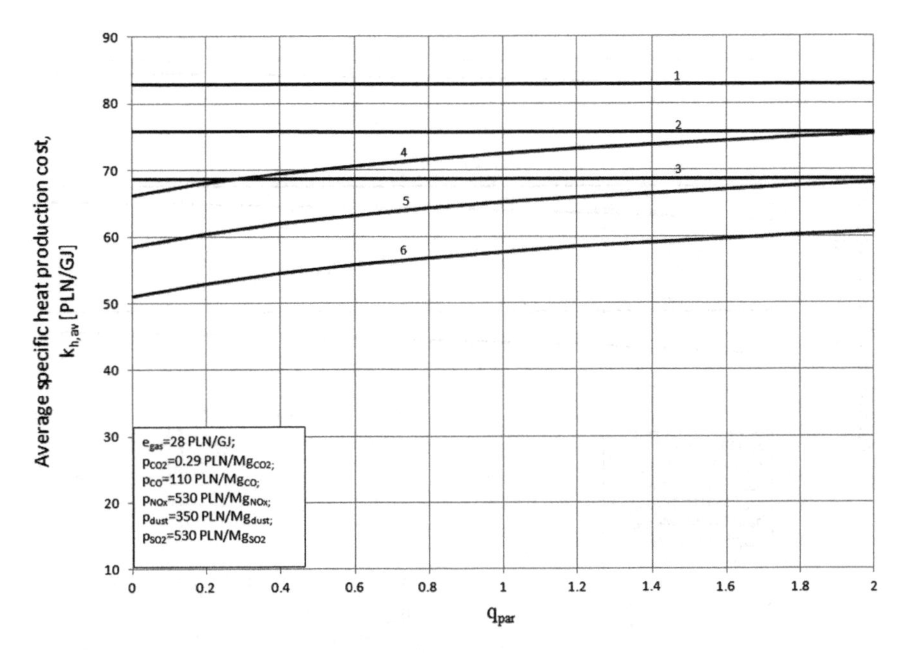

Fig. 4.9 Average specific cost of heat production as a function of q_{par} with the purchase price of carbon dioxide emission allowances as a parameter, where: 1 applied to $e_{CO_2} = 100$ PLN/Mg, $e_{el} =$ 170 PLN/MWh, $e^{bo}_{coal} = 19.64$ PLN/GJ; 2 applied to $e_{CO_2} = 60$ PLN/Mg, $e_{el} = 170$ PLN/MWh, e^{bo}_{coal} $= 19.86$ PLN/GJ; 3 applied to $e_{CO_2} = 20$ PLN/Mg, $e_{el} = 170$ PLN/MWh, $e^{bo}_{coal} = 20.08$ PLN/GJ; 4 applied to $e_{CO_2} = 100$ PLN/Mg, $e_{el} = 170$ PLN/MWh; $e_{coal} = 11.4$ PLN/GJ; 5 applied to $e_{CO_2} =$ 60 PLN/Mg, $e_{el} = 170$ PLN/MWh; $e_{coal} = 11.4$ PLN/GJ; 6 applied to $e_{CO_2} = 20$ PLN/Mg, $e_{el} =$ 170 PLN/MWh; $e_{coal} = 11.4$ PLN/GJ

$$
\begin{aligned}
&+ \rho^{coal}_{dust} p^{t=0}_{dust} \frac{1}{a_{dust} - r}\left[e^{(a_{dust}-r)T} - 1\right] + (1-u)\rho^{coal}_{CO_2} e^{t=0}_{CO_2} \frac{1}{b_{CO_2} - r}\left[e^{(b_{CO_2}-r)T} - 1\right]\Bigg\} \\
&+ \frac{i}{\tau_s}(1 - e^{-rT})(1 + x_{sal,t,ins})\frac{\delta_{serv}}{r} + \frac{zi}{\tau_s}\left(\frac{1 - e^{-rT}}{T} + 1\right) \\
&- \frac{[q_{par}(1-\eta_{GT})\eta_{HRSG} + \eta_B]\eta_{SH}\eta^k_{ST}\eta_{me} + q_{par}\eta_{GT}}{\beta[q_{par}(1-\eta_{GT})\eta_{HRSG} + \eta_B]\eta_{SH}\eta^k_{ST}\eta_{HE}}(1-\varepsilon_{el})e^{t=0}_{el}\frac{1}{a_{el} - r}\left[e^{(a_{el}-r)T} - 1\right]\Bigg\}. \quad (4.23)
\end{aligned}
$$

This formula was obtained from Eq. (3.5) after substituting dependencies resulting from the energy balance presented in Fig. 3.5:

$$
\frac{E^{gas}_{ch,A}}{Q_A} = \frac{q_{par}}{\beta[q_{par}(1-\eta_{GT})\eta_{HRSG} + \eta_B]\eta_{SH}\eta^k_{ST}\eta_{HE}}, \quad (4.24)
$$

$$
\frac{E^{coal}_{ch,A}}{Q_A} = \frac{1}{\beta[q_{par}(1-\eta_{GT})\eta_{HRSG} + \eta_B]\eta_{SH}\eta^k_{ST}\eta_{HE}}, \quad (4.25)
$$

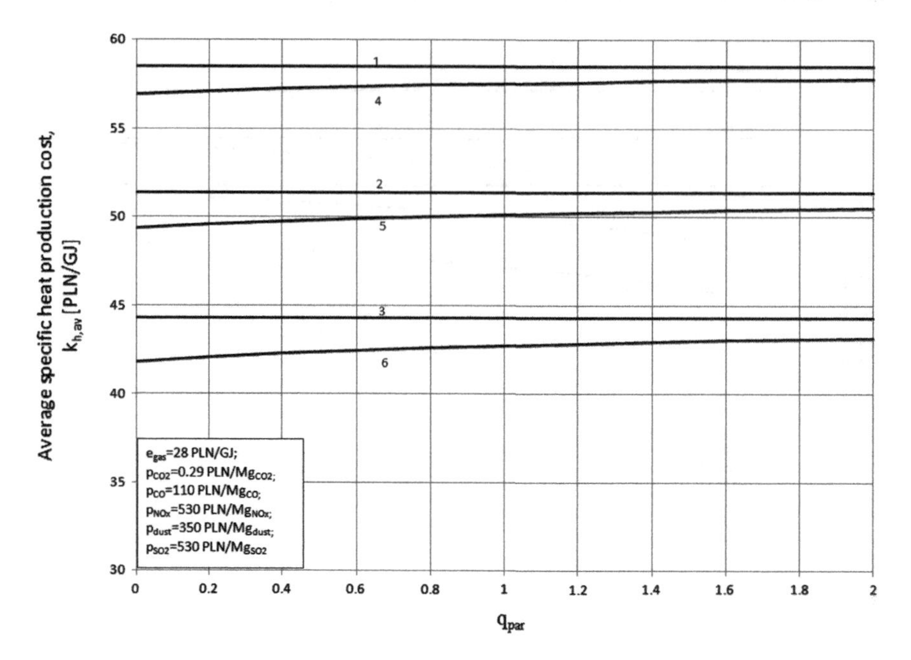

Fig. 4.9a Average specific cost of heat production as a function of q_{par} with the purchase price of carbon dioxide emission allowances as a parameter, where: 1 applied to $e_{CO_2} = 100$ PLN/Mg, $e_{el} =$ 220 PLN/MWh, $e^{bo}_{coal} = 12.17$ PLN/GJ; 2 applied to $e_{CO_2} = 60$ PLN/Mg, $e_{el} = 220$ PLN/MWh, e^{bo}_{coal} = 12.38 PLN/GJ; 3 applied to $e_{CO_2} = 20$ PLN/Mg, $e_{el} = 220$ PLN/MWh, $e^{bo}_{coal} = 12.61$ PLN/GJ; 4 applied to $e_{CO_2} = 100$ PLN/Mg, $e_{el} = 220$ PLN/MWh; $e_{coal} = 11.4$ PLN/GJ; 5 applied to $e_{CO_2} =$ 60 PLN/Mg, $e_{el} = 220$ PLN/MWh; $e_{coal} = 11.4$ PLN/GJ; 6 applied to $e_{CO_2} = 20$ PLN/Mg, $e_{el} =$ 220 PLN/MWh; $e_{coal} = 11.4$ PLN/GJ

$$\frac{E^{ST}_{el,A} + E^{GT}_{el,A}}{Q_A} = \frac{[q_{par}(1 - \eta_{GT})\eta_{HRSG} + \eta_B]\eta_{SH}\eta^k_{ST}\eta_{me} + q_{par}\eta_{GT}}{\beta[q_{par}(1 - \eta_{GT})\eta_{HRSG} + \eta_B]\eta_{SH}\eta^k_{ST}\eta_{HE}}. \tag{4.26}$$

When calculating from the formula (4.23) derivative $dk_{h,av}/dq_{par}$, the relationship is obtained

$$\begin{aligned}
\frac{dk_{h,av}}{dq_{par}} &= \frac{r}{1 - e^{-rT}} \Bigg\{ \frac{\beta\eta_B\eta_{SH}\eta^k_{ST}\eta_{HE}}{\left\{\beta[q_{par}(1 - \eta_{GT})\eta_{HRSG} + \eta_B]\eta_{SH}\eta^k_{ST}\eta_{HE}\right\}^2} \\
&\times \Bigg\{(1 + x_{sw,m,was})e^{t=0}_{gas}\frac{1}{a_{gas} - r}[e^{(a_{gas}-r)T} - 1] \\
&+ \rho^{gas}_{CO_2}p^{t=0}_{CO_2}\frac{1}{a_{CO_2} - r}[e^{(a_{CO_2}-r)T} - 1] + \rho^{gas}_{CO}p^{t=0}_{CO}\frac{1}{a_{CO} - r}[e^{(a_{CO}-r)T} - 1] \\
&+ \rho^{gas}_{NO_X}p^{t=0}_{NO_X}\frac{1}{a_{NO_X} - r}[e^{(a_{NO_X}-r)T} - 1] + \rho^{gas}_{SO_2}p^{t=0}_{SO_2}\frac{1}{a_{SO_2} - r}[e^{(a_{SO_2}-r)T} - 1] \\
&+ \rho^{gas}_{dust}p^{t=0}_{dust}\frac{1}{a_{dust} - r}[e^{(a_{dust}-r)T} - 1] + (1 - u)\rho^{gas}_{CO_2}e^{t=0}_{CO_2}\frac{1}{b_{CO_2} - r}[e^{(b_{CO_2}-r)T} - 1]\Bigg\}
\end{aligned}$$

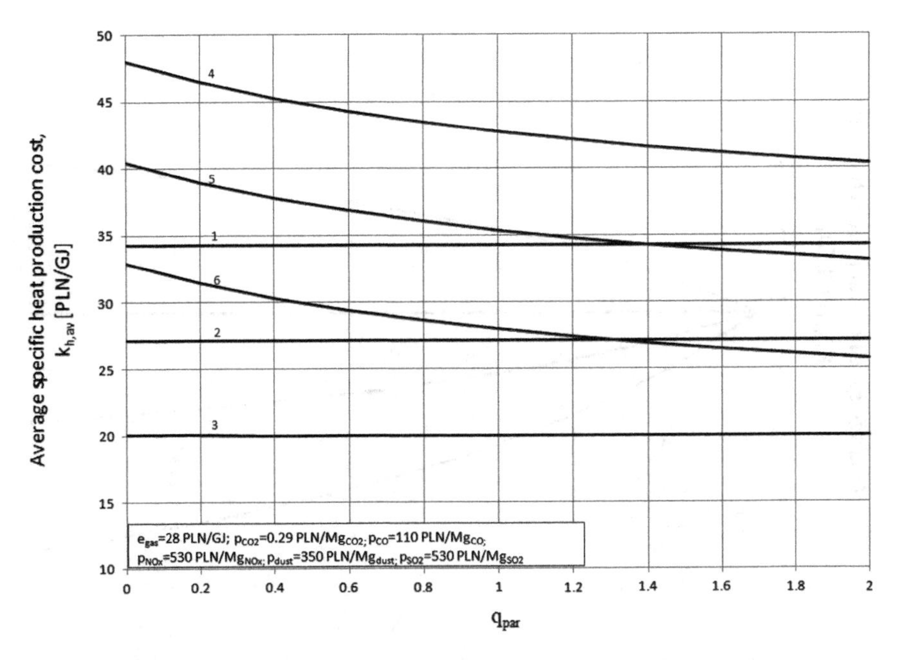

Fig. 4.9b Average specific cost of heat production as a function of q_{par} with the purchase price of carbon dioxide emission allowances as a parameter, where: 1″ applied to $e_{co_2} = 100$ PLN/Mg, $e_{el} = 270$ PLN/MWh, $e^{bo}_{coal} = 4.72$ PLN/GJ; 2″ applied to $e_{co_2} = 60$ PLN/Mg, $e_{el} = 270$ PLN/MWh, $e^{bo}_{coal} = 4.94$ PLN/GJ; 3″ applied to $e_{co_2} = 20$ PLN/Mg, $e_{el} = 270$ PLN/MWh, $e^{bo}_{coal} = 5.17$ PLN/GJ; 4″ applied to $e_{co_2} = 100$ PLN/Mg, $e_{el} = 270$ PLN/MWh; $e_{coal} = 11.4$ PLN/GJ; 5″ applied to $e_{co_2} = 60$ PLN/Mg, $e_{el} = 270$ PLN/MWh; $e_{coal} = 11.4$ PLN/GJ; 6″ applied to $e_{co_2} = 20$ PLN/Mg, $e_{el} = 270$ PLN/MWh; $e_{coal} = 11.4$ PLN/GJ

$$
\begin{aligned}
&- \frac{\beta(1-\eta_{GT})\eta_{HRSG}\eta_{SH}\eta^{k}_{ST}\eta_{HE}}{\left\{\beta[q_{par}(1-\eta_{GT})\eta_{HRSG}+\eta_B]\eta_{SH}\eta^{k}_{ST}\eta_{HE}\right\}^2} \\
&\times \Bigg\{(1+x_{sw,m,was})e^{t=0}_{coal}\frac{1}{a_{coal}-r}[\mathrm{e}^{(a_{coal}-r)T}-1] \\
&+ \rho^{coal}_{\mathrm{CO_2}}p^{t=0}_{\mathrm{CO_2}}\frac{1}{a_{\mathrm{CO_2}}-r}[\mathrm{e}^{(a_{\mathrm{CO_2}}-r)T}-1] + \rho^{coal}_{\mathrm{CO}}p^{t=0}_{\mathrm{CO}}\frac{1}{a_{\mathrm{CO}}-r}[\mathrm{e}^{(a_{\mathrm{CO}}-r)T}-1] \\
&+ \rho^{coal}_{\mathrm{NO_X}}p^{t=0}_{\mathrm{NO_X}}\frac{1}{a_{\mathrm{NO_X}}-r}[\mathrm{e}^{(a_{\mathrm{NO_X}}-r)T}-1] + \rho^{coal}_{\mathrm{SO_2}}p^{t=0}_{\mathrm{SO_2}}\frac{1}{a_{\mathrm{SO_2}}-r}[\mathrm{e}^{(a_{\mathrm{SO_2}}-r)T}-1] \\
&+ \rho^{coal}_{dust}p^{t=0}_{dust}\frac{1}{a_{dust}-r}[\mathrm{e}^{(a_{dust}-r)T}-1] + (1-u)\rho^{coal}_{\mathrm{CO_2}}e^{t=0}_{\mathrm{CO_2}}\frac{1}{b_{\mathrm{CO_2}}-r}[\mathrm{e}^{(b_{\mathrm{CO_2}}-r)T}-1]\Bigg\} \\
&+ \Bigg\{\frac{\beta(1-\eta_{GT})\eta_{HRSG}\eta_{SH}\eta^{k}_{ST}\eta_{HE}\left\{[q_{par}(1-\eta_{GT})\eta_{HRSG}+\eta_B]\eta_{SH}\eta^{k}_{ST}\eta_{me}+q_{par}\eta_{GT}\right\}}{\left\{\beta[q_{par}(1-\eta_{GT})\eta_{HRSG}+\eta_B]\eta_{SH}\eta^{k}_{ST}\eta_{HE}\right\}^2} \\
&- \frac{[(1-\eta_{GT})\eta_{HRSG}\eta_{SH}\eta^{k}_{ST}\eta_{me}+\eta_{GT}][q_{par}(1-\eta_{GT})\eta_{HRSG}+\eta_B]\beta\eta_{SH}\eta^{k}_{ST}\eta_{HE}}{\left\{\beta[q_{par}(1-\eta_{GT})\eta_{HRSG}+\eta_B]\eta_{SH}\eta^{k}_{ST}\eta_{HE}\right\}^2}\Bigg\} \\
&\times (1-\varepsilon_{el})e^{t=0}_{el}\frac{1}{a_{el}-r}[\mathrm{e}^{(a_{el}-r)T}-1]\Bigg\}
\end{aligned}
\tag{4.27}
$$

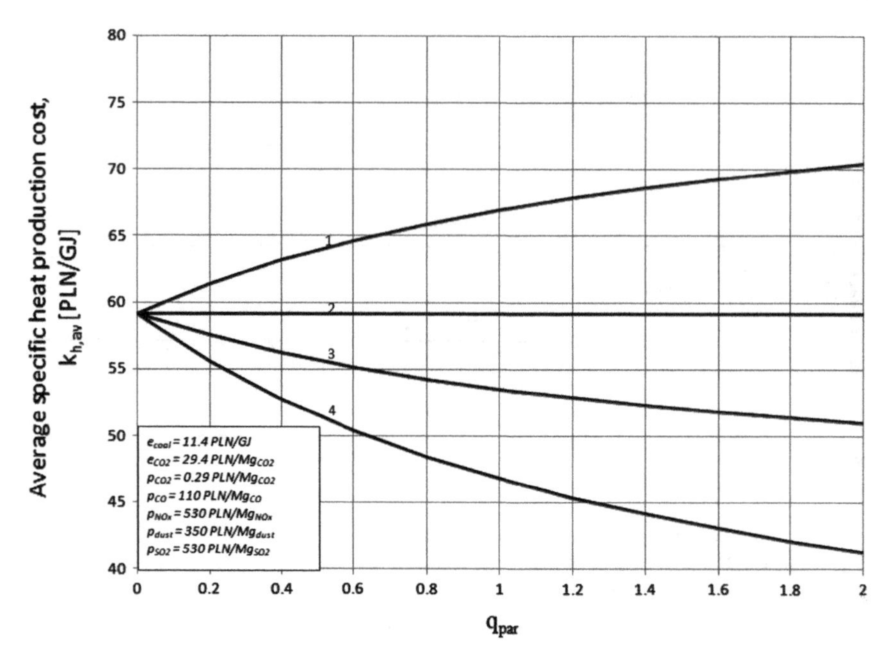

Fig. 4.10 Average specific cost of heat production as a function of q_{par} with the price of gas as a parameter, where: 1 applied to e_{gas} = 32 PLN/GJ, e_{el} = 170 PLN/MWh; 2 applied to e^{bo}_{gas} = 22.71 PLN/GJ; e_{el} = 170 PLN/MWh; 3 applied to e_{gas} = 16 PLN/GJ, e_{el} = 170 PLN/MWh; 4 applied to e_{gas} = 8 PLN/GJ, e_{el} = 170 PLN/MWh

and from this dependence of necessary condition on the existence of the cost extreme $k_{h,av}$, i.e. the resetting of the derivative $dk_{h,av}/dq_{par} = 0$, an identical formula is obtained as the formula (4.20). This is a consequence of dependence lack adoption in the mathematical model of specific investment outlays "i" on the variable q_{par}. The boundary gas price in the case of its combustion in a combined heat and power plant—Figs. 4.10, 4.10a, 4.10b, 4.11, 4.11a and 4.11b—for given coal and electricity prices does not depend on the type of steam turbine installed in it, despite the fact that the specific cost $k_{h,av}$ obviously depends on it. This is because the specific outlays are greater for a combined heat and power plant with extraction-condensing steam turbine than for extraction-backpressure steam turbine. This results from the necessity of incurring investment outlays on the steam condenser and the cooling tower in the system with extraction-condensing turbine, and such an expenditures do not exist in the system with extraction-backpressure turbine.

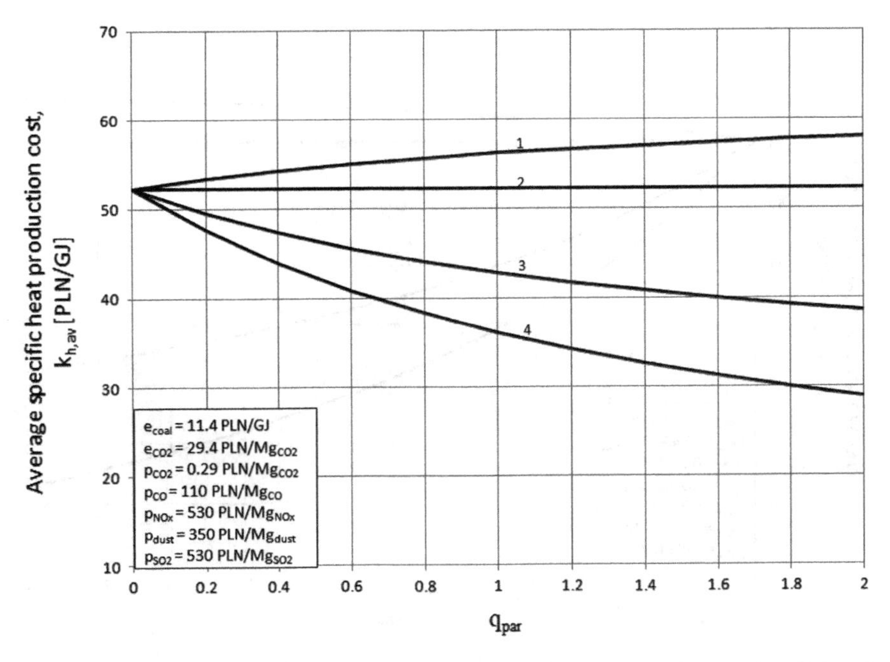

Fig. 4.10a Average specific cost of heat production as a function of q_{par} with the price of gas as a parameter, where: 1 $e_{gas} = 32$ PLN/GJ, $e_{el} = 220$ PLN/MWh; 2 $e_{gas}^{bo} = 27.29$ PLN/GJ; $e_{el} = 220$ PLN/MWh; 3 $e_{gas} = 16$ PLN/GJ, $e_{el} = 220$ PLN/MWh; 4 $e_{gas} = 8$ PLN/GJ, $e_{el} = 220$ PLN/MWh

4.3.1 Discussion and Analysis of Calculation Results

In Figs. 4.10, 4.10a and 4.10b, the curve courses 3 and 4 are decreasing because gas costs in Eq. (4.27) are lower than costs associated with coal and then $dk_{h,av}/dq_{par} < 0$. The values of these curves are calculated for gas prices lower than its limit prices equal to: $e_{gas}^{bo} = 22.71$ PLN/GJ (for $e_{el} = 170$ PLN/MWh), $e_{gas}^{bo} = 27.29$ PLN/GJ (for $e_{el} = 220$ PLN/MWh), $e_{gas}^{bo} = 31.86$ PLN/GJ (for $e_{el} = 270$ PLN/MWh). The gas limit prices correspond to the current price of coal equal to $e_{coal} = 11.4$ PLN/GJ and current purchase price of CO_2 emission allowances equal to $e_{CO_2} =$ PLN 29.4/Mg. Curves 1 in Figs. 4.10, 4.10a and 4.10b, are of increasing nature, because the price of gas is higher for them than the boundary prices, and therefore the costs related to gas in Eq. (4.27) exceed the costs associated with coal and $dk_{h,av}/dq_{par} > 0$. As can be seen from the course of curves 3 and 4 dual-fuel gas-steam CHP plant is more economically profitable than a single coal-fired CHP plant if the gas price is lower than its boundary prices set for current prices coal and electricity.

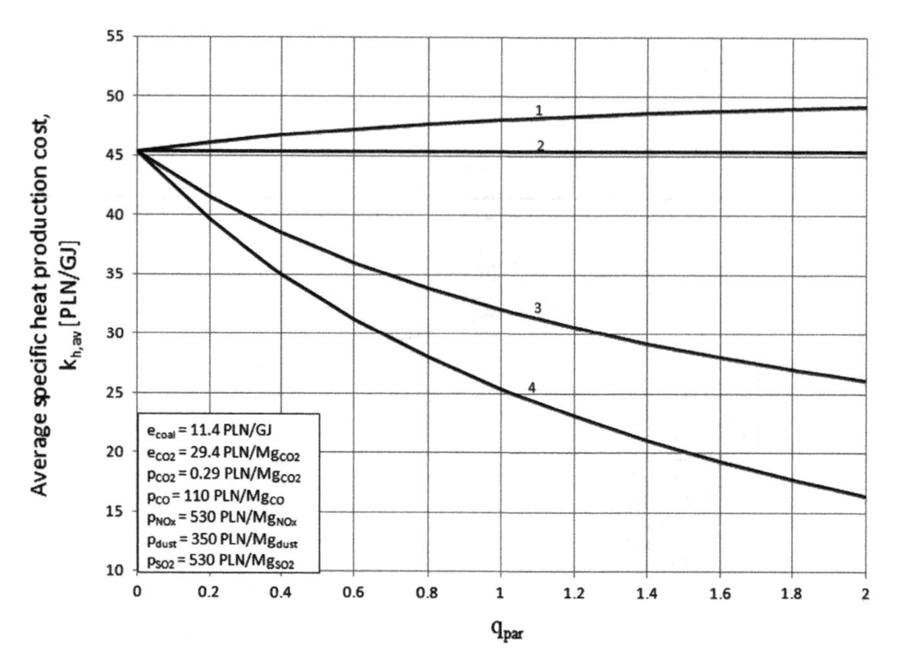

Fig. 4.10b Average specific cost of heat production as a function of q_{par} with the price of gas as a parameter, where: 1 e_{gas} = 35 PLN/GJ, e_{el} = 270 PLN/MWh, 2 e^{bo}_{gas} = 31.86 PLN/GJ; e_{el} = 270 PLN/MWh; 3 e_{gas} = 16 PLN/GJ, e_{el} = 270 PLN/MWh, 4 e_{gas} = 8 PLN/GJ, e_{el} = 270 PLN/MWh

The specific heat production cost $k_{h,av}$ is then the smallest for the maximum power of the gas turbine set. For current coal e_{coal} = 11.4 PLN/GJ and electricity prices e_{el} = 170 PLN/MWh, the price of gas must be lower than the price e^{bo}_{gas} = 22.71 PLN/GJ. The current gas price in Poland is equal to e_{gas} = 28 PLN/GJ and dual-fuel gas-steam combined heat and power plants are therefore unprofitable.

The curve 3 in Figs. 4.11, 4.11a and 4.11b. have an increasing course, because the gas costs in Eq. (4.27) outweigh the costs associated with coal. The values of these curves are calculated for prices of coal e_{coal} lower than its boundary prices equal to: e^{bo}_{coal} = 20 PLN/GJ (for = 170 PLN/MWh), e^{bo}_{coal} = 12.55 PLN/GJ (for e_{el} = 220 PLN/MWh), e^{bo}_{coal} = 5.11 PLN/GJ (for e_{el} = 270 PLN/MWh) (curves 2, 2′, 2″). The boundary prices of coal correspond to the current price of gas equal to e_{gas} = 28 PLN/GJ and current price for the purchase of CO_2 emission allowances e_{CO_2} = 29.4 PLN/Mg. On the other hand, the curve 1 in that Figures have a decreasing character because they were made for the price of coal higher than the boundary ones, and therefore the relation $dk_{h,av}/dq_{par} < 0$ is fulfilled. The cost of heat production $k_{h,av}$ is also the highest due to the highest price of coal and high gas price.

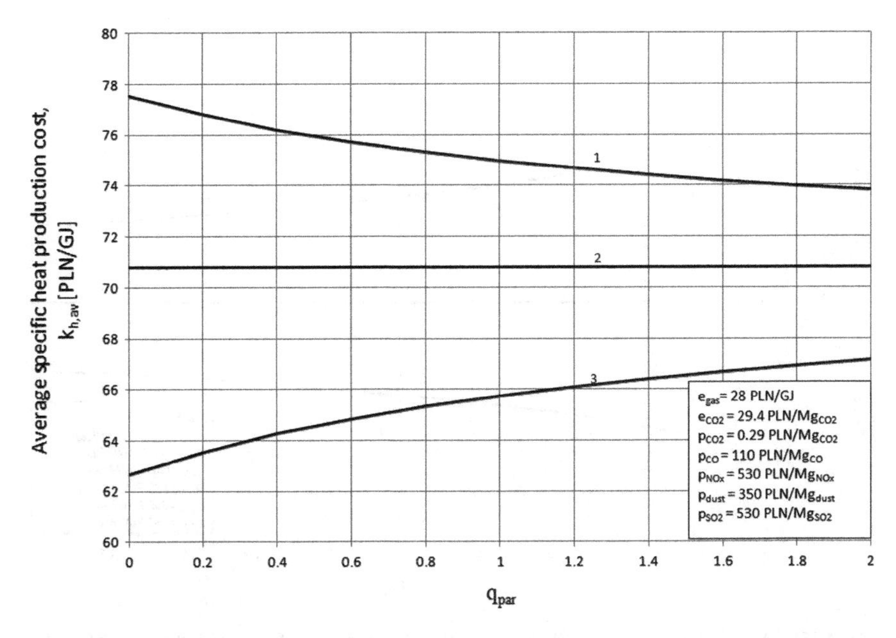

Fig. 4.11 Average specific cost of heat production as a function of q_{par} with the price of coal as a parameter, where: 1 applied to e_{coal} = 25 PLN/GJ, e_{el} = 170 PLN/MWh; 2 applied to e^{bo}_{coal} = 20 PLN/GJ, e_{el} = 170 PLN/MWh; 3 applied to e_{coal} = 14 PLN/GJ, e_{el} = 170 PLN/MWh

Figures 4.12 and 4.12a shows the impact of the purchase price of e_{CO_2} allowances for a tonne of carbon dioxide emission on the specific heat production cost $k_{h,av}$. The higher the e_{CO_2} price, the higher the e^{bo}_{coal} boundary value, so the higher the price of gas for which the construction of dual-fuel gas-steam CHP plant will be economically profitable. This does not mean, however, that increasing the price e_{CO_2} is beneficial. It is the opposite. Increasing the e_{CO_2} price causes an increase in the $k_{h,av}$ cost.

The curves 1, 2, 3 in Fig. 4.12a. are of increasing nature for the same reason as the curves in Figs. 4.10, 4.10a, 4.10b, 4.11, 4.11a and 4.11b. The values of these curves were calculated for the price of coal e_{coal} = 11.4 PLN/GJ lower than its boundary prices equal to: e^{bo}_{coal} = 19.64 PLN/GJ (for e_{CO_2} = 100 PLN/Mg, e_{el} = 170 PLN/MWh), e^{bo}_{coal} = 19.86 PLN/GJ (for e_{CO_2} = 60 PLN/Mg, e_{el} = 170 PLN/MWh), e^{bo}_{coal} = 20.08 PLN/GJ (for e_{CO_2} = 20 PLN/Mg, e_{el} = 170 PLN/MWh) (curves 1, 2, 3 in Fig. 4.12). All boundary prices of coal correspond to the price of gas equal to e_{gas} = 28 PLN/GJ.

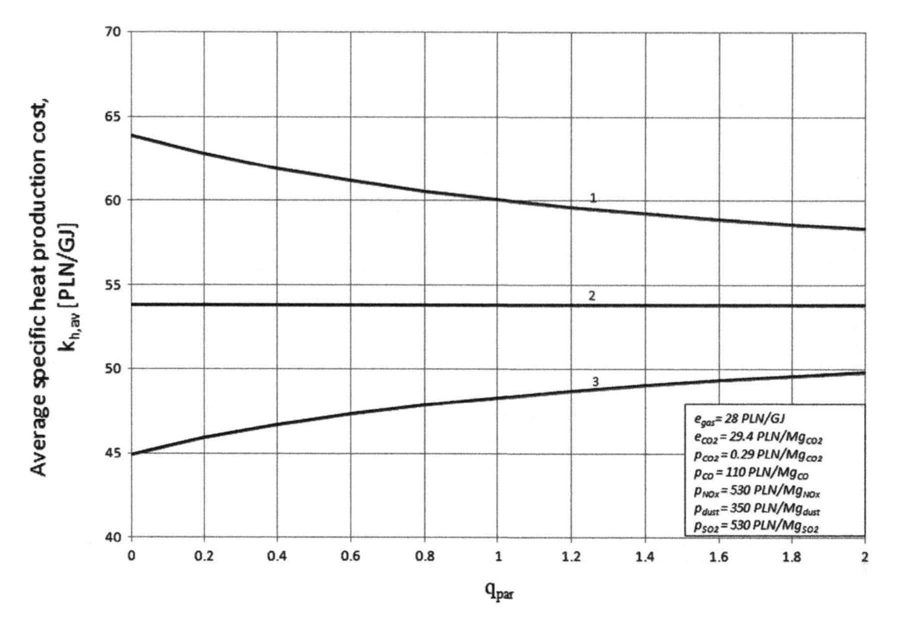

Fig. 4.11a Average specific cost of heat production as a function of q_{par} with the price of coal as a parameter, where: 1 applied to e_{coal} = 20 PLN/GJ, e_{el} = 220 PLN/MWh; 2 applied to e^{bo}_{coal} = 12.55 PLN/GJ; e_{el} = 220 PLN/MWh; 3 applied to e_{coal} = 6 PLN/GJ, e_{el} = 220 PLN/MWh

Figures 4.13, 4.13a and 4.13b shows the gas e^{bo}_{gas} and coal e^{bo}_{coal} boundary prices calculated using formula (4.20), i.e. prices for which the cost $k_{h,av}$ assumes a constant value independent of q_{par}. The higher the boundary coal price, the higher the boundary gas price, and vice versa. As it results from the necessary condition (4.21), the higher the price e^{bo}_{gas}, the higher the price e_{gas} for which the construction of dual-fuel gas-steam CHP plant is profitable.

Figure 4.14 presents a wide range of the boundary value of gas price e^{bo}_{gas} as a function of the boundary price of coal e^{bo}_{coal} with electricity e_{el} and allowances for emission of a ton of carbon dioxide e_{CO_2} prices as parameters.

As can be seen from the course of the curves in Fig. 4.14, the boundary gas and coal prices for dual-fuel gas-steam CHP plants depend to a large extent on the e_{el} price of electricity. The purchase price e_{CO_2} of permits for the emission of tonne of CO_2 affects these prices to a negligible extent.

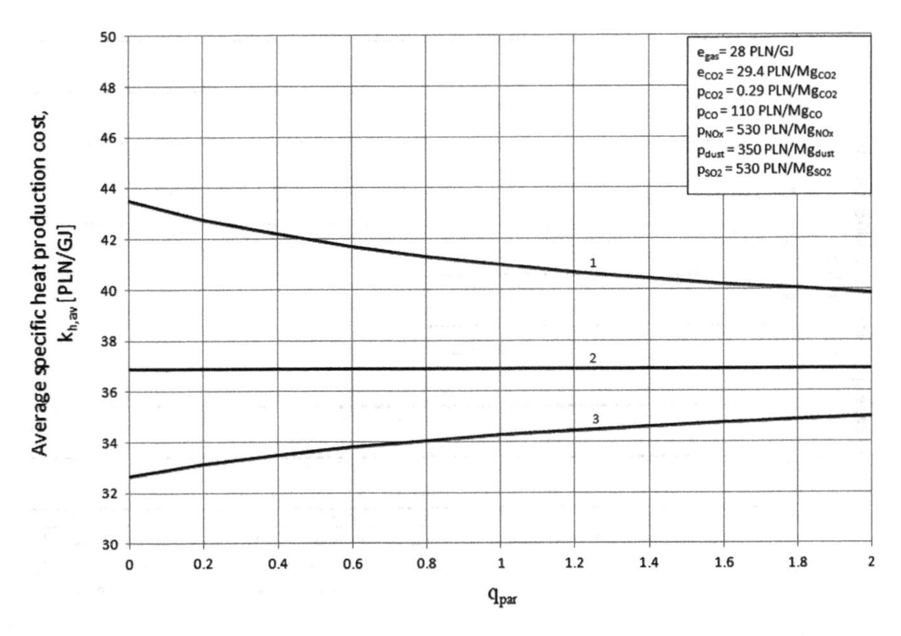

Fig. 4.11b Average specific cost of heat production as a function of q_{par} with the price of coal as a parameter, where: 1 applied to $e_{coal} = 10$ PLN/GJ, $e_{el} = 270$ PLN/MWh; 2 applied to $e^{bo}_{coal} = 5.11$ PLN/GJ, $e_{el} = 270$ PLN/MWh; 3 applied to $e_{coal} = 2$ PLN/GJ, $e_{el} = 270$ PLN/MWh

4.3.2 Summary

Using the methodologies and mathematical models presented in the paper for the technical and economic analysis of the operation of the CHP plant in dual-fuel gas technology in a parallel system, multivariate calculations were performed, which are presented in Figs. 4.7, 4.7a, 4.7b, 4.8, 4.8a, 4.8b, 4.9, 4.9a, 4.9b, 4.10, 4.10a, 4.10b, 4.11, 4.11a, 4.11b, 4.12, 4.12a, 4.13, 4.13a, 4.13b and 4.14. They are presented for a combined heat and power plant with extraction-condensing steam turbine and extraction backpressure steam turbine for a wide range of changes in energy prices and environmental charges, and therefore also for current prices and environmental fees.

The decisive values of the power of the gas turbine set $N^{GT}_{el\,\max}$ are the prices of gas, coal, electricity, environmental costs and capital expenditures. The fuel prices and outlays are lower, and the higher the price of electricity, the lower the specific cost of heat production $k_{h,av}$ formula (3.5).

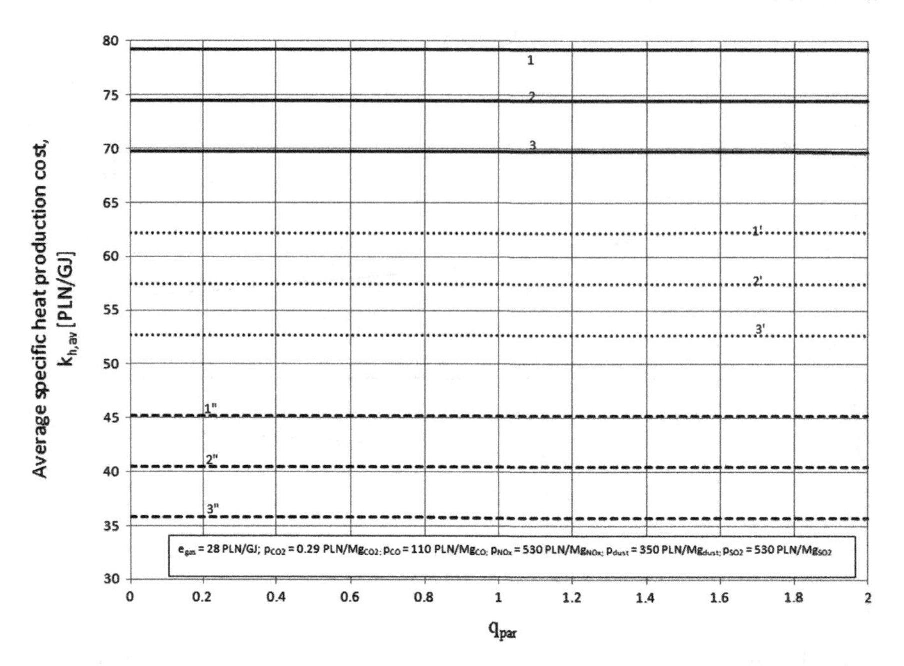

Fig. 4.12 Average specific cost of heat production as a function of q_{par} with the price of the purchase of CO_2 emission allowances e_{CO_2} as a parameter, where: 1 applied to $e_{co_2} = 100$ PLN/Mg, $e_{el} = 170$ PLN/MWh, $e^{bo}_{coal} = 19.64$ PLN/GJ; 1′ applied to $e_{co_2} = 100$ PLN/Mg, $e_{el} = 220$ PLN/MWh, $e^{bo}_{coal} = 12.17$ PLN/GJ; 1″ applied to $e_{co_2} = 100$ PLN/Mg, $e_{el} = 270$ PLN/MWh, $e^{bo}_{coal} = 4.72$ PLN/GJ; 2 applied to $e_{co_2} = 60$ PLN/Mg, $e_{el} = 170$ PLN/MWh, $e^{bo}_{coal} = 19.86$ PLN/GJ; 2′ applied to $e_{co_2} = 60$ PLN/Mg, $e_{el} = 220$ PLN/MWh, $e^{bo}_{coal} = 12.38$ PLN/GJ; 2″ applied to $e_{co_2} = 60$ PLN/Mg, $e_{el} = 270$ PLN/MWh, $e^{bo}_{coal} = 4.94$ PLN/GJ; 3 applied to $e_{co_2} = 20$ PLN/Mg, $e_{el} = 170$ PLN/MWh, $e^{bo}_{coal} = 20.08$ PLN/GJ; 3′ applied to $e_{co_2} = 20$ PLN/Mg, $e_{el} = 220$ PLN/MWh, $e^{bo}_{coal} = 12.61$ PLN/GJ; 3″ applied to $e_{co_2} = 20$ PLN/Mg, $e_{el} = 270$ PLN/MWh, $e^{bo}_{coal} = 5.17$ PLN/GJ

A *necessary condition* for the cost-effectiveness of construction of dual-fuel gas-steam CHP plants, and hence their greater profitability from single coal CHP plants, is the fulfillment of the relationship $e_{gas} \leq e^{bo}_{gas}$ formula (4.21), i.e. that the gas price would be lower than its boundary price. What is more, the lower the price of gas than the boundary price (determined for current coal and electricity prices), the lower is the specific cost of heat production $k_{h,av}$ and the most advantageous is the maximum power of the gas turbine set $N^{GT}_{el\,\max}$-curves 3 and 4 in Figs. 4.7, 4.7a, 4.7b, 4.10, 4.10a and 4.10b.

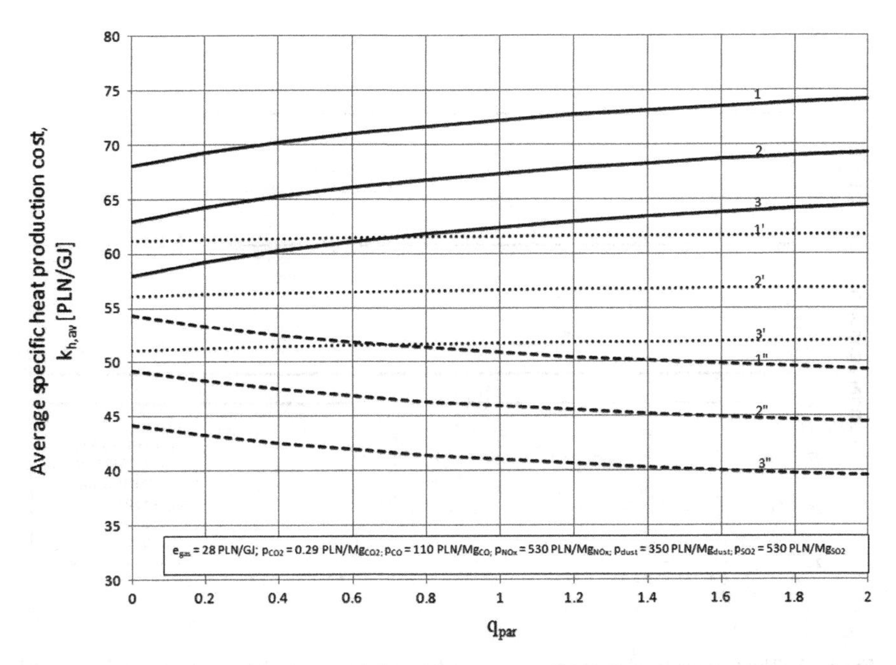

Fig. 4.12a Average specific cost of heat production as a function of q_{par} with the price of the purchase of CO_2 emission allowances e_{CO_2} as a parameter, where: 1 applied to $e_{co_2} = 100$ PLN/Mg, $e_{el} = 170$ PLN/MWh; $e_{coal} = 11.4$ PLN/GJ; 1′ applied to $e_{co_2} = 100$ PLN/Mg, $e_{el} = 220$ PLN/MWh; $e_{coal} = 11.4$ PLN/GJ; 1″ applied to $e_{co_2} = 100$ PLN/Mg, $e_{el} = 270$ PLN/MWh; $e_{coal} = 11.4$ PLN/GJ; 2 applied to $e_{co_2} = 60$ PLN/Mg, $e_{el} = 170$ PLN/MWh; $e_{coal} = 11.4$ PLN/GJ; 2′ applied to $e_{co_2} = 60$ PLN/Mg, $e_{el} = 220$ PLN/MWh; $e_{coal} = 11.4$ PLN/GJ; 2″ applied to $e_{co_2} = 60$ PLN/Mg, $e_{el} = 270$ PLN/MWh; $e_{coal} = 11.4$ PLN/GJ; 3 applied to $e_{co_2} = 20$ PLN/Mg, $e_{el} = 170$ PLN/MWh; $e_{coal} = 11.4$ PLN/GJ; 3′ applied to $e_{co_2} = 20$ PLN/Mg, $e_{el} = 220$ PLN/MWh; $e_{coal} = 11.4$ PLN/GJ; 3″ applied to $e_{co_2} = 20$ PLN/Mg, $e_{el} = 270$ PLN/MWh; $e_{coal} = 11.4$ PLN/GJ

The higher the purchase price e_{CO_2} of CO_2 emission permits (Figs. 4.9, 4.9a 4.12, 4.12a, and 4.12b), the higher the boundary values e^{bo}_{coal} and e^{bo}_{gas} (Fig. 4.14), and thus the price of gas for which the construction of dual-fuel gas-steam CHP plant is profitable may be higher. However, this does not mean, as already stated, that increasing the price e_{CO_2} is beneficial. It is absolutely the opposite, because increasing the price e_{CO_2}, which the European Union aims at, increases the cost of heat production $k_{h,av}$, thus drains the pockets of its users to a greater extent.

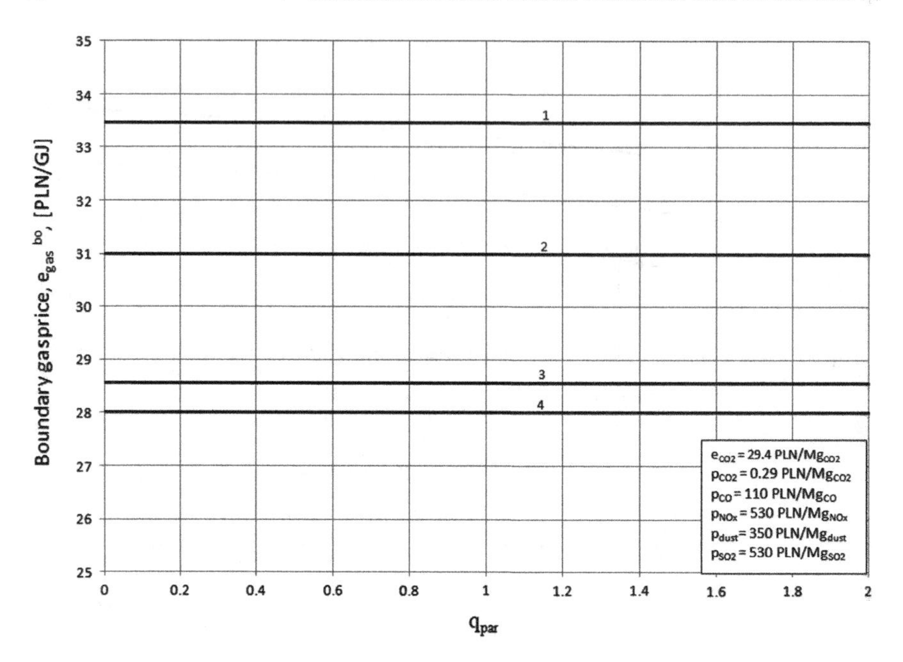

Fig. 4.13 The boundary price of gas e^{bo}_{gas} in the function of q_{par} with the boundary price of coal e^{bo}_{coal} as a parameter where: 1 applied to $e^{bo}_{coal} = 14$ PLN/GJ, $e_{el} = 270$ PLN/MWh; 2 applied to $e^{bo}_{coal} = 10$ PLN/GJ, $e_{el} = 270$ PLN/MWh; 3 applied to $e^{bo}_{coal} = 6$ PLN/GJ, $e_{el} = 270$ PLN/MWh; 4 applied to $e^{bo}_{coal} = 5.11$ PLN/GJ, $e_{el} = 270$ PLN/MWh

The combined heat and power plant with an extraction-backpressure turbine is more economically advantageous due to the lower heat production costs $k_{h,av}$. It should be noted, however, that systems with extraction- backpressure turbine are only valid for industrial combined heat and power stations, i.e. when the heating demand for technological processes is constant throughout the year. The systems with the extraction-condensing turbine, on the other hand, allow for a year-round flexible operation of the combined heat and power plant, i.e. work independent of the changing needs for heat. In surplus heating steam, it is supplied to the condensation

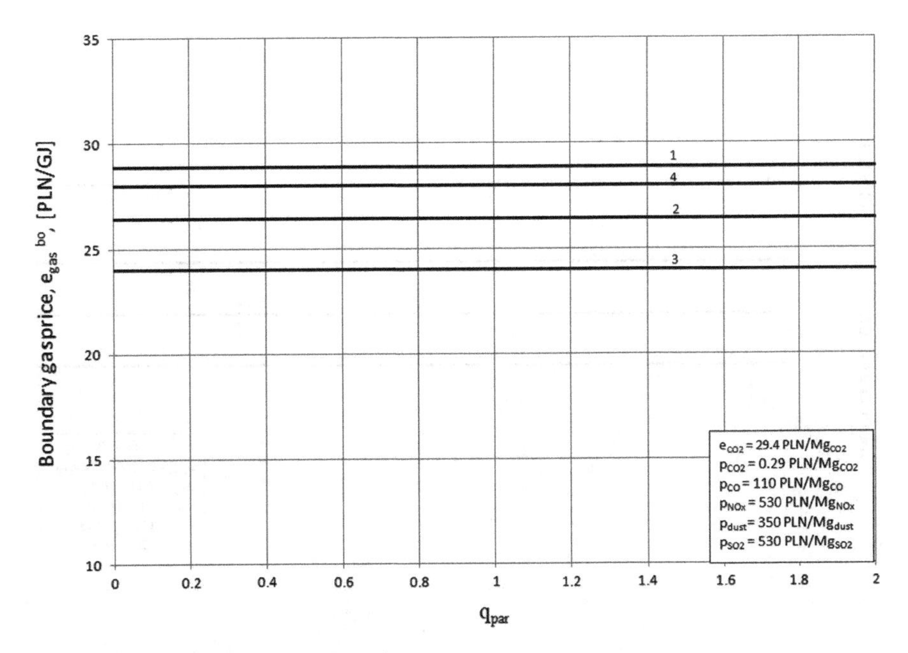

Fig. 4.13a The boundary price of gas e_{gas}^{bo} in the function of q_{par} with the boundary price of coal e_{coal}^{bo} as a parameter where: 1 applied to $e_{coal}^{bo} = 14$ PLN/GJ; $e_{el} = 220$ PLN/MWh; 2 applied to $e_{coal}^{bo} = 10$ PLN/GJ; $e_{el} = 220$ PLN/MWh; 3 applied to $e_{coal}^{bo} = 6$ PLN/GJ; $e_{el} = 220$ PLN/MWh; 4 applied to $e_{coal}^{bo} = 12.55$ PLN/GJ; $e_{el} = 220$ PLN/MWh

turbine set and it generates electric energy, the most noble form of energy. On the other hand, the use of a backpressure turbine would cause that in the case of reduced thermal needs, the gas turbine would have to work, for example, partly on the "bypass stack" (outlet of exhaust from the turbine into the stack instead of into the recovery boiler) or with an incomplete load. The system operation would then be thermodynamically and economically ineffective. Other possibilities could be the operation of a steam boiler with variable capacity, which, however, significantly increases the technical consumption of the boiler, or turn off block operation and start-up the operation

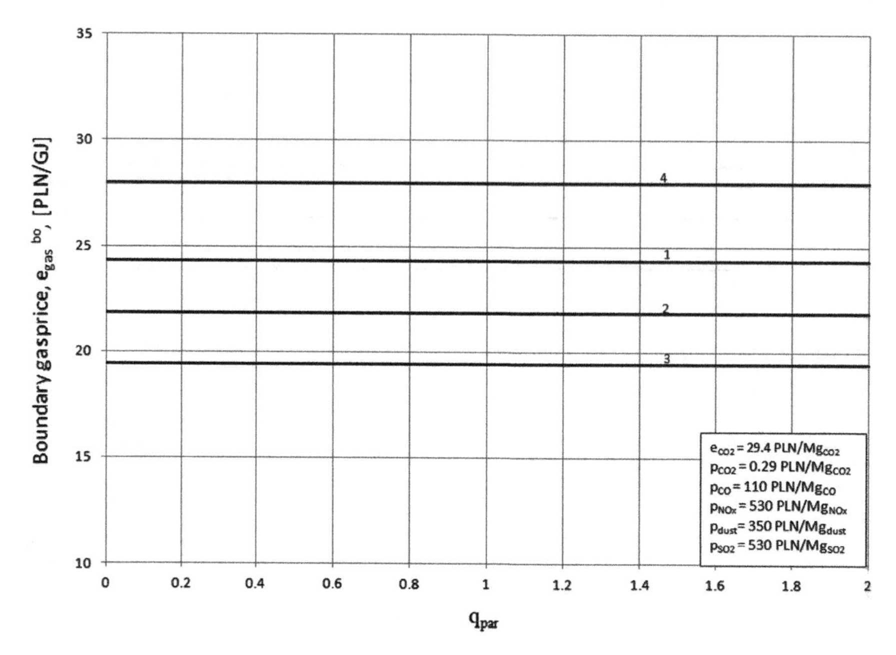

Fig. 4.13b The boundary price of gas e_{gas}^{bo} in the function of q_{par} with the boundary price of coal e_{coal}^{bo} as a parameter where: 1 applied to $e_{coal}^{bo} = 14$ PLN/GJ; $e_{el} = 170$ PLN/MWh; 2 applied to $e_{coal}^{bo} = 10$ PLN/GJ; $e_{el} = 170$ PLN/MWh; 3 applied to $e_{coal}^{bo} = 6$ PLN/GJ; $e_{el} = 170$ PLN/MWh; 4 applied to $e_{coal}^{bo} = 20$ PLN/GJ; $e_{el} = 170$ PLN/MWh

of reserve-peak water boilers in the combined heat and power plant. However, the condition of "paying back" the investment is the work of basic equipment. When the devices are not working or they are short, not fully used, they may not only "not bring" profit, but also not return the incurred investment outlays. Therefore, in municipal heat and power plants, the installation of a steam-operated extraction-condensing turbine allows, as mentioned above, for a very flexible year-round operation of the system, for adapting to the variable demand for heat and operation even with total condensation—provided there are no restrictions on the electricity sale.

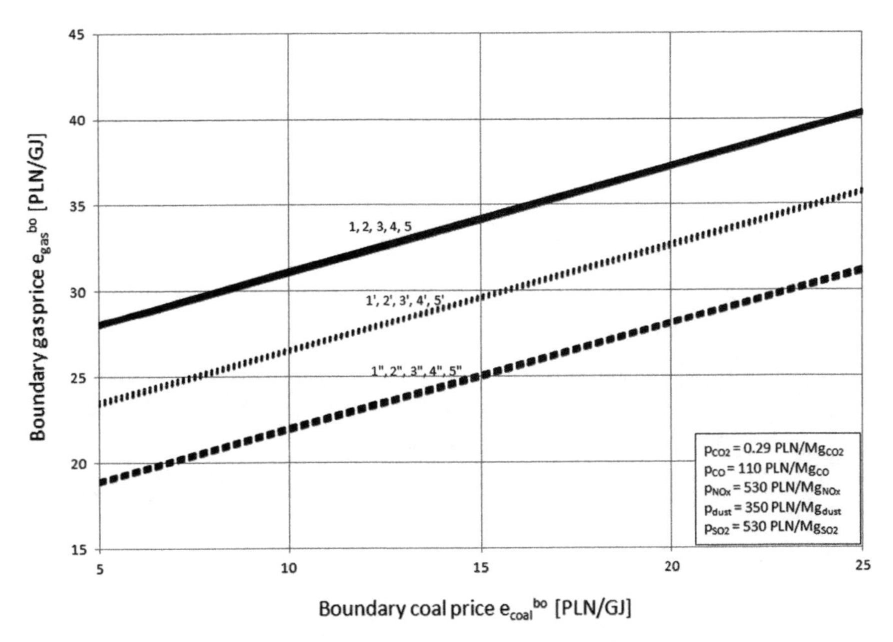

Fig. 4.14 The boundary price of gas e_{gas}^{bo} as a function of the boundary price of coal e_{coal}^{bo} with the price of CO_2 emissions e_{CO_2} as a parameter, where: 1 applied to $e_{CO_2} = 100$ PLN/Mg, $e_{el} = 270$ PLN/MWh; 2 applied to $e_{CO_2} = 80$ PLN/Mg, $e_{el} = 270$ PLN/MWh; 3 applied to $e_{CO_2} = 60$ PLN/Mg, $e_{el} = 270$ PLN/MWh; 4 applied to $e_{CO_2} = 40$ PLN/Mg, $e_{el} = 270$ PLN/MWh; 5 applied to $e_{CO_2} = 20$ PLN/Mg, $e_{el} = 270$ PLN/MWh; $1'$ applied to $e_{CO_2} = 100$ PLN/Mg, $e_{el} = 220$ PLN/MWh; $2'$ applied to $e_{CO_2} = 80$ PLN/Mg, $e_{el} = 220$ PLN/MWh; $3'$ applied to $e_{CO_2} = 60$ PLN/Mg, $e_{el} = 220$ PLN/MWh; $4'$ applied to $e_{CO_2} = 40$ PLN/Mg, $e_{el} = 220$ PLN/MWh; $5'$ applied to $e_{CO_2} = 20$ PLN/Mg, $e_{el} = 220$ PLN/MWh; $1''$ applied to $e_{CO_2} = 100$ PLN/Mg, $e_{el} = 170$ PLN/MWh; $2''$ applied to $e_{CO_2} = 80$ PLN/Mg, $e_{el} = 170$ PLN/MWh; $3''$ applied to $e_{CO_2} = 60$ PLN/Mg, $e_{el} = 170$ PLN/MWh; $4''$ applied to $e_{CO_2} = 40$ PLN/Mg, $e_{el} = 170$ PLN/MWh; $5''$ applied to $e_{CO2} = 20$ PLN/Mg, $e_{el} = 170$ PLN/MWh

References

1. Bartnik R, Bartnik B, Hnydiuk-Stefan A (2016) optimum investment strategy in the power industry Mathematical Models. Springer, New York
2. Bartnik R, Buryn Z, Hnydiuk-Stefan A (2017) Investment strategy in heating and CHP Mathematical Models. Wydawnictwo. Springer, London

Chapter 5
Methodology and Mathematical Models with Continuous Time for Technical and Economic Analysis of Effectiveness Modernization of Existing Coal Blocks for Dual-Fuel Gas-Steam Systems

One of the possible directions of modernization of technological systems of coal-fired power plants and combined heat and power plants may be improvement of already existing in their structure individual installations and devices, modernization of boilers increasing their energy efficiency, modernization of turbine flow systems increasing their internal efficiency, modernization of cooling systems or treatments reducing electrical own needs of power plants and combined heat and power plants. Such treatments are not able to significantly improve the efficiency of electricity and heat generation in them, since they do not change the thermal parameters of the Clausius-Rankine heating circle. It can be expected then to improve the energy efficiency of the power plant and combined heat and power plant with a maximum of 1–3 percentage points.

The technologically and technically rational way of modernization of already existing coal-fired power plants and CHP plants, which makes them modern, is their conversion to dual-fuel gas-steam systems powered with hard coal and natural gas.

Such a conversion is possible by their integration with modern gas technologies based on gas turbines powered by natural gas [1–4]. The thermal circuit implemented in them will change at that time. In addition to Clausius-Rankine's current steam turbine cycle, the Joule gas turbine cycle will be implemented in them, which will result in a very significant improvement in their energy efficiency. The use of gas turbines in coal systems creates many possibilities of solutions for the construction of new, dual-fuel, thermal structures of these systems.

A. Hnydiuk-Stefan, *Dual-Fuel Gas-Steam Power Block Analysis*, Power Systems,
https://doi.org/10.1007/978-3-030-03050-6_5

The repowering of thermal systems of power plants and combined heat and power plants provides, in principle, almost unlimited possibilities for solutions of these systems. This is due not only to the diversity and complexity of the coal structures of power plants and combined heat and power plants themselves, in which there may be at the same time water boilers, steam boilers, backpressure steam turbines, bleed-backpressure turbines, extraction-condensing turbines, but also from the way of modernization.

In general, it can be concluded that the unlimited number of solutions results:

from the diversity and complexity of the structures of coal thermal systems in power plants and combined heat and power plant
from the method of repowering of a coal system with a gas turbine
from the power used in the gas turbine system, regardless of how this repowering was implemented.

Thermodynamic analysis should therefore be carried out for specific power plants and combined heat and power plants individually. It is possible, however, to generalize the economic analysis of these systems, i.e. it is possible to generalize the formulas for the economic viability of modernization of coal-fired power plants and coal-fired CHP plants. The structure of these formulas is the same for all solutions. The difference in the analysis of specific systems will consist only in substituting for these formulas other values of individual components, such as: investment outlays, amount of fuel burned, amount of electricity produced in the system, amount of harmful combustion products emitted into the environment, etc. The selection methodology presented in this chapter for optimal thermal structures for requiring modernization of coal-fired power plants and combined heat and power plants [formulas (5.1)–(5.4)] have the same general character, and cover all possible solutions.

The unexpressed *expressis verbis* of the above facts shows the need to develop methodologies and mathematical models describing the functional space of technical and economic phenomena occurring in the processes of electric energy and heat production in order to analyze them in modernized sources for dual-fuel gas-steam systems. This chapter therefore presents this methodology and models. The following (Fig. 5.1) is a time diagram used to build them.

The time intervals $\langle 0, t_1 \rangle$, $\langle t_1, t_2 \rangle$, $\langle t_2, T \rangle$ represent the years of operation of the power plant and heat and power plant before, during and after its modernization. When, in the models presented below, in the time period $\langle 0, t_1 \rangle$, the value of zero for electricity production is taken into account, they relate to modernized heating plants.

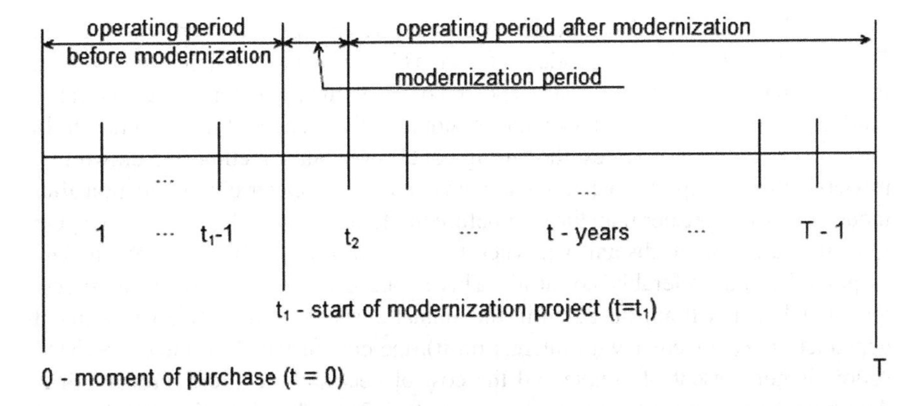

Fig. 5.1 Time diagram of the work of a modernized power plant, combined heat and power plant and heating plant

5.1 Methodology with Continuous Time for Technical and Economic Analysis of Modernization Coal-Fired Power Plants and Coal-Fired CHP Plants for Dual-Fuel Gas-Steam Systems

The fundamental dependence by which the technical and economic efficiency of the modernization of the power plant, CHP plant and heating plant is determined, including capital expenditure J_M for this modernization is the equation for the total *NPV* profit that is achieved from their operation for a period of T years:

$$
\begin{aligned}
NPV = & \int_{0}^{t_1} [F + A + (S - K_e - F - A)(1 - p)]\mathrm{e}^{-rt} dt \\
& + \int_{t_1}^{t_2} \left[F + A + F^M + A^M + \left(S^M - K_e^M - F - A - F^M - A^M\right)(1 - p)\right]\mathrm{e}^{-rt} dt \\
& + \int_{t_2}^{T} \left[F + A + F^M + A^M + \left(S^{\mathrm{mod}} - K_e^{\mathrm{mod}} - F - A - F^M - A^M\right)(1 - p)\right]\mathrm{e}^{-rt} dt \\
& - \int_{0}^{T} (F + R)\mathrm{e}^{-rt} dt - \int_{t_1}^{T} \left(F^M + R^M\right)\mathrm{e}^{-rt} dt \rightarrow \max.
\end{aligned}
\tag{5.1}
$$

In the formula (5.1), the same designations were used as in the formula (2.21) in Sect. 2.1. With the help of dependence formula (5.1), it is also possible to analyze the impact of moments t_1, t_2 of the start and end of modernization of heat and electricity production sources—Fig. 3.1—on the economic effectiveness of their operation. In the case of amortized sources, the wrong conclusion can sometimes be drawn that modernization is unprofitable, as the unit cost of heat and electricity production after modernization is greater than the cost before modernization, and is smaller, the year t_1 of its start is more distant, i.e. when the modernization will take place as late as possible, and preferably not at all. This is because, in the modernized source, the capital cost will appear again in the annual costs of its operation (installment depreciation J_M together with interest on it), the cost of which in the depreciated source is not present of course and the cost of heat and electricity production is determined solely by costs exploitation [formula (2.26)]. Therefore, the annual cost of generating heat and electricity in the depreciated source is low. However, it should be remembered that if the source is not modernized, its further use may be impossible, it will have to be switched off due to technical wear. Hence the need for modernization.

5.2 Mathematical Models with the Continuous Time of Technical and Economic Analysis of the Modernization of Coal-Fired Power Plants and Coal-Fired Heaters for Dual-Fuel Gas-Steam Systems

Integrating the Eq. (5.1) while maintaining the same designations and scenarios of changes in operating and capital costs as well as revenues as in Sect. 2.1, the equation for the total *NPV* profit obtained from the work of the modernized CHP plant is obtained:

$$
\begin{aligned}
NPV = & \Bigg\{ \Bigg\{ E_{el,A} e_{el}^{t=0} \frac{1}{a_{el} - r} \left[e^{(a_{el} - r)t_1} - 1 \right] + Q_A e_h^{t=0} \frac{1}{a_h - r} [e^{(a_h - r)t_1} - 1] \\
& - \frac{E_{el,A} + Q_A}{\eta_{CHP}} (1 + x_{sw,m,was}) e_{fuel}^{t=0} \frac{1}{a_{fuel} - r} \left[e^{(a_{fuel} - r)t_1} - 1 \right] \\
& - \frac{E_{el,A} + Q_A}{\eta_{CHP}} \rho_{CO_2} p_{CO_2}^{t=0} \frac{1}{a_{CO_2} - r} \left[e^{(a_{CO_2} - r)t_1} - 1 \right] \\
& - \frac{E_{el,A} + Q_A}{\eta_{CHP}} \rho_{CO} p_{CO}^{t=0} \frac{1}{a_{CO} - r} \left[e^{(a_{CO} - r)t_1} - 1 \right] \\
& - \frac{E_{el,A} + Q_A}{\eta_{CHP}} \rho_{NO_X} p_{NO_X}^{t=0} \frac{1}{a_{NO_X} - r} \left[e^{(a_{NO_X} - r)t_1} - 1 \right] \\
& - \frac{E_{el,A} + Q_A}{\eta_{CHP}} \rho_{SO_2} p_{SO_2}^{t=0} \frac{1}{a_{SO_2} - r} \left[e^{(a_{SO_2} - r)t_1} - 1 \right] \\
& - \frac{E_{el,A} + Q_A}{\eta_{CHP}} \rho_{dust} p_{dust}^{t=0} \frac{1}{a_{dust} - r} \left[e^{(a_{dust} - r)t_1} - 1 \right] \\
& - \frac{E_{el,A} + Q_A}{\eta_{CHP}} (1 - u) \rho_{CO_2} e_{CO_2}^{t=0} \frac{1}{b_{CO_2} - r} \left[e^{(b_{CO_2} - r)t_1} - 1 \right] \\
& - J(1 + x_{sal,t,ins}) \frac{\delta_{serv}}{r} \left(1 - e^{-rt_1}\right) - J_0 \left[1 + \frac{1}{T} - \left(1 + \frac{1}{T} - \frac{t_1}{T}\right) e^{-rt_1} \right] \Bigg\} \\
& + \Bigg\{ E_{el,A}^M e_{el}^{M,t=t_1} \frac{1}{a_{el}^M - r} \left[e^{(a_{el}^M - r)t_2} - e^{(a_{el}^M - r)t_1} \right] \\
& + Q_A^M e_h^{t=t_1} \frac{1}{a_h^M - r} \left[e^{(a_h^M - r)t_2} - e^{(a_h^M - r)t_1} \right] \\
& - \frac{E_{el,A}^M + Q_A^M}{\eta_{CHP}^M} (1 + x_{sw,m,was}) e_{coal}^{M,t=t_1} \frac{1}{a_{coal}^M - r} \left[e^{(a_{coal}^M - r)t_2} - e^{(a_{coal}^M - r)t_1} \right] \\
& - \frac{E_{el,A}^M + Q_A^M}{\eta_{CHP}^M} \rho_{CO_2} p_{CO_2}^{M,t=t_1} \frac{1}{a_{CO_2}^M - r} \left[e^{\left(a_{CO_2}^M - r\right)t_2} - e^{\left(a_{CO_2}^M - r\right)t_1} \right] \\
& - \frac{E_{el,A}^M + Q_A^M}{\eta_{CHP}^M} \rho_{CO} p_{CO}^{M,t=t_1} \frac{1}{a_{CO}^M - r} \left[e^{(a_{CO}^M - r)t_2} - e^{(a_{CO}^M - r)t_1} \right] \\
& - \frac{E_{el,A}^M + Q_A^M}{\eta_{CHP}^M} \rho_{NO_X} p_{NO_X}^{M,t=t_1} \frac{1}{a_{NO_X}^M - r} \left[e^{\left(a_{NO_X}^M - r\right)t_2} - e^{\left(a_{NO_X}^M - r\right)t_1} \right] \\
& - \frac{E_{el,A}^M + Q_A^M}{\eta_{CHP}^M} \rho_{SO_2} p_{SO_2}^{M,t=t_1} \frac{1}{a_{SO_2}^M - r} \left[e^{\left(a_{SO_2}^M - r\right)t_2} - e^{\left(a_{SO_2}^M - r\right)t_1} \right] \\
& - \frac{E_{el,A}^M + Q_A^M}{\eta_{CHP}^M} \rho_{dust} p_{dust}^{M,t=t_1} \frac{1}{a_{dust}^M - r} \left[e^{(a_{dust}^M - r)t_1} - e^{(a_{dust}^M - r)t_1} \right] \\
& - \frac{E_{el,A}^M + Q_A^M}{\eta_{CHP}^M} \left(1 - u^M\right) \rho_{CO_2} e_{CO_2}^{M,t=t_1} \frac{1}{b_{CO_2}^M - r} \left[e^{\left(b_{CO_2}^M - r\right)t_2} - e^{\left(b_{CO_2}^M - r\right)t_1} \right] \\
& - \frac{(J + J_M)(1 + x_{sal,t,ins}) \delta_{serv}^M}{r} \left(e^{-rt_1} - e^{-rt_2}\right) \\
& - J_0 \left[\left(1 + \frac{1}{T} - \frac{t_1}{T}\right) e^{-rt_1} - \left(1 + \frac{1}{T} - \frac{t_2}{T}\right) e^{-rt_2} \right]
\end{aligned}
$$

$$-J_M\left[\left(1+\frac{1}{T-t_1}-\frac{t_1}{T-t_1}\right)\mathrm{e}^{-rt_1}-\left(1+\frac{1}{T-t_1}-\frac{t_2}{T-t_1}\right)\mathrm{e}^{-rt_2}\right]\Bigg\}$$

$$+\Bigg\{E_{el,A}^{\,mod}\,\mathrm{e}_{el}^{mod,t=t_2}\frac{1}{a_{el}^{\,mod}-r}\left[\mathrm{e}^{(a_{el}^{\,mod}-r)T}-\mathrm{e}^{(a_{el}^{\,mod}-r)t_2}\right]$$

$$+Q_A^{\,mod}\,\mathrm{e}_h^{t=t_2}\frac{1}{a_h^{\,mod}-r}[\mathrm{e}^{(a_h^{\,mod}-r)T}-\mathrm{e}^{(a_h^{\,mod}-r)t_2}]$$

$$-\frac{E_{el,A}^{\,mod}+Q_A^{\,mod}}{\eta_{CHP}^{\,mod}}(1+x_{sw,m,was})\mathrm{e}_{coal}^{mod,t=t_2}\frac{1}{a_{coal}^{\,mod}-r}\left[\mathrm{e}^{(a_{coal}^{\,mod}-r)T}-\mathrm{e}^{(a_{coal}^{\,mod}-r)t_2}\right]$$

$$-\frac{E_{el,A}^{\,mod}+Q_A^{\,mod}}{\eta_{CHP}^{\,mod}}\rho_{CO_2}p_{CO_2}^{mod,t=t_2}\frac{1}{a_{CO_2}^{\,mod}-r}\left[\mathrm{e}^{\left(a_{CO_2}^{\,mod}-r\right)T}-\mathrm{e}^{\left(a_{CO_2}^{\,mod}-r\right)t_2}\right]$$

$$-\frac{E_{el,A}^{\,mod}+Q_A^{\,mod}}{\eta_{CHP}^{\,mod}}\rho_{CO}\,p_{CO}^{mod,t=t_2}\frac{1}{a_{CO}^{\,mod}-r}\left[\mathrm{e}^{(a_{CO}^{\,mod}-r)T}-\mathrm{e}^{(a_{CO}^{\,mod}-r)t_2}\right]$$

$$-\frac{E_{el,A}^{\,mod}+Q_A^{\,mod}}{\eta_{CHP}^{\,mod}}\rho_{NO_X}p_{NO_X}^{mod,t=t_2}\frac{1}{a_{NO_X}^{\,mod}-r}\left[\mathrm{e}^{\left(a_{NO_X}^{\,mod}-r\right)T}-\mathrm{e}^{\left(a_{NO_X}^{\,mod}-r\right)t_2}\right]$$

$$-\frac{E_{el,A}^{\,mod}+Q_A^{\,mod}}{\eta_{CHP}^{\,mod}}\rho_{SO_2}p_{SO_2}^{mod,t=t_2}\frac{1}{a_{SO_2}^{\,mod}-r}\left[\mathrm{e}^{\left(a_{SO_2}^{\,mod}-r\right)T}-\mathrm{e}^{\left(a_{SO_2}^{\,mod}-r\right)t_2}\right]$$

$$-\frac{E_{el,A}^{\,mod}+Q_A^{\,mod}}{\eta_{CHP}^{\,mod}}\rho_{dust}\,p_{dust}^{mod,t=t_2}\frac{1}{a_{dust}^{\,mod}-r}\left[\mathrm{e}^{(a_{dust}^{\,mod}-r)T}-\mathrm{e}^{(a_{dust}^{\,mod}-r)t_2}\right]$$

$$-\frac{E_{el,A}^{\,mod}+Q_A^{\,mod}}{\eta_{CHP}^{\,mod}}\left(1-u^{\,mod}\right)\rho_{CO_2}\mathrm{e}_{CO_2}^{mod,t=t_2}\frac{1}{b_{CO_2}^{\,mod}-r}\left[\mathrm{e}^{\left(b_{CO_2}^{\,mod}-r\right)T}-\mathrm{e}^{\left(b_{CO_2}^{\,mod}-r\right)t_2}\right]$$

$$-\frac{(J+J_M)(1+x_{sal,t,ins})\delta_{serv}^{\,mod}}{r}\left(\mathrm{e}^{-rt_2}-\mathrm{e}^{-rT}\right)-J_0\left[\left(1+\frac{1}{T}-\frac{t_2}{T}\right)\mathrm{e}^{-rt_2}-\frac{1}{T}\mathrm{e}^{-rT}\right]$$

$$-J_M\left[\left(1+\frac{1}{T-t_1}-\frac{t_2}{(T-t_1)}\right)\mathrm{e}^{-rt_2}-\left(1+\frac{1}{T-t_1}-\frac{T}{(T-t_1)}\right)\mathrm{e}^{-rT}\right]\Bigg\}\Bigg\}(1-p) \tag{5.2}$$

J_0 means discounted for the moment of completion of the construction of a power plant, heat and power plant or heat plant, expenditures J incurred for their construction in the period b years of its duration, $J_0 = Jz$ [formula (2.17)].

In the case of the analysis of the operation of the power plant in the formula (3.2), the zero value should be used as heating, $Q_A = 0$, in the case of heating plant analysis, the zero value should be substituted for electricity production, $E_{el,A} = 0$.

Equivalent to the $NPV \to \max$ criterion for searching for the optimal technology for modernization of combined heat and power plants and heating plants is the criterion of minimizing the average unit heat production cost in them over the T years of their operation. This cost is determined from formula (5.2) from the condition $NPV = 0$ at $a_h = a_h^M = a_h^{\text{mod}} = 0$ [in the case of a power plant, the unit cost of electricity production is expressed by the formula (5.4)]:

$$
\begin{aligned}
k_{h,av} = \Bigg\{ & -E_{el,A} e_{el}^{t=0} \frac{1}{a_{el} - r}\left[e^{(a_{el}-r)t_1} - 1\right] \\
& + \frac{E_{el,A} + Q_A}{\eta_{CHP}}(1 + x_{sw,m,was}) e_{coal}^{t=0} \frac{1}{a_{coal} - r}\left[e^{(a_{coal}-r)t_1} - 1\right] \\
& + \frac{E_{el,A} + Q_A}{\eta_{CHP}} \rho_{CO_2} p_{CO_2}^{t=0} \frac{1}{a_{CO_2} - r}\left[e^{(a_{CO_2}-r)t_1} - 1\right] \\
& + \frac{E_{el,A} + Q_A}{\eta_{CHP}} \rho_{CO} p_{CO}^{t=0} \frac{1}{a_{CO} - r}\left[e^{(a_{CO}-r)t_1} - 1\right] \\
& + \frac{E_{el,A} + Q_A}{\eta_{CHP}} \rho_{NO_X} p_{NO_X}^{t=0} \frac{1}{a_{NO_X} - r}\left[e^{(a_{NO_X}-r)t_1} - 1\right] \\
& + \frac{E_{el,A} + Q_A}{\eta_{CHP}} \rho_{SO_2} p_{SO_2}^{t=0} \frac{1}{a_{SO_2} - r}\left[e^{(a_{SO_2}-r)t_1} - 1\right] \\
& + \frac{E_{el,A} + Q_A}{\eta_{CHP}} \rho_{dust} p_{dust}^{t=0} \frac{1}{a_{dust} - r}\left[e^{(a_{dust}-r)t_1} - 1\right] \\
& + \frac{E_{el,A} + Q_A}{\eta_{CHP}} (1 - u) \rho_{CO_2} e_{CO_2}^{t=0} \frac{1}{b_{CO_2} - r}\left[e^{(b_{CO_2}-r)t_1} - 1\right] \\
& + J(1 + x_{sal,t,ins}) \frac{\delta_{serv}}{r}\left(1 - e^{-rt_1}\right) + J_0\left[1 + \frac{1}{T} - \left(1 + \frac{1}{T} - \frac{t_1}{T}\right)e^{-rt_1}\right] \\
& - E_{el,A}^M e_{el}^{M,t=t_1} \frac{1}{a_{el}^M - r}\left[e^{(a_{el}^M-r)t_2} - e^{(a_{el}^M-r)t_1}\right] \\
& + \frac{E_{el,A}^M + Q_A^M}{\eta_{CHP}^M}(1 + x_{sw,m,was}) e_{coal}^{M,t=t_1} \frac{1}{a_{coal}^M - r}\left[e^{(a_{coal}^M-r)t_2} - e^{(a_{coal}^M-r)t_1}\right] \\
& + \frac{E_{el,A}^M + Q_A^M}{\eta_{CHP}^M} \rho_{CO_2} p_{CO_2}^{M,t=t_1} \frac{1}{a_{CO_2}^M - r}\left[e^{\left(a_{CO_2}^M-r\right)t_2} - e^{\left(a_{CO_2}^M-r\right)t_1}\right] \\
& + \frac{E_{el,A}^M + Q_A^M}{\eta_{CHP}^M} \rho_{CO} p_{CO}^{M,t=t_1} \frac{1}{a_{CO}^M - r}\left[e^{(a_{CO}^M-r)t_2} - e^{(a_{CO}^M-r)t_1}\right] \\
& + \frac{E_{el,A}^M + Q_A^M}{\eta_{CHP}^M} \rho_{NO_X} p_{NO_X}^{M,t=t_1} \frac{1}{a_{NO_X}^M - r}\left[e^{\left(a_{NO_X}^M-r\right)t_2} - e^{\left(a_{NO_X}^M-r\right)t_1}\right] \\
& + \frac{E_{el,A}^M + Q_A^M}{\eta_{CHP}^M} \rho_{SO_2} p_{SO_2}^{M,t=t_1} \frac{1}{a_{SO_2}^M - r}\left[e^{\left(a_{SO_2}^M-r\right)t_2} - e^{\left(a_{SO_2}^M-r\right)t_1}\right] \\
& + \frac{E_{el,A}^M + Q_A^M}{\eta_{CHP}^M} \rho_{dust} p_{dust}^{M,t=t_1} \frac{1}{a_{dust}^M - r}\left[e^{(a_{dust}^M-r)t_1} - e^{(a_{dust}^M-r)t_1}\right] \\
& + \frac{E_{el,A}^M + Q_A^M}{\eta_{CHP}^M}\left(1 - u^M\right) \rho_{CO_2} e_{CO_2}^{M,t=t_1} \frac{1}{b_{CO_2}^M - r}\left[e^{\left(b_{CO_2}^M-r\right)t_2} - e^{\left(b_{CO_2}^M-r\right)t_1}\right]
\end{aligned}
$$

$$
\begin{aligned}
&+\frac{(J+J_M)(1+x_{sal,t,ins})\delta_{serv}^{M}}{r}\left(\mathrm{e}^{-rt_1}-\mathrm{e}^{-rt_2}\right)\\
&+J_0\left[\left(1+\frac{1}{T}-\frac{t_1}{T}\right)\mathrm{e}^{-rt_1}-\left(1+\frac{1}{T}-\frac{t_2}{T}\right)\mathrm{e}^{-rt_2}\right]\\
&+J_M\left[\left(1+\frac{1}{T-t_1}-\frac{t_1}{T-t_1}\right)\mathrm{e}^{-rt_1}-\left(1+\frac{1}{T-t_1}-\frac{t_2}{T-t_1}\right)\mathrm{e}^{-rt_2}\right]\\
&-E_{el,A}^{\,\mathrm{mod}}\,\mathrm{e}_{el}^{\mathrm{mod},t=t_2}\frac{1}{a_{el}^{\mathrm{mod}}-r}\left[\mathrm{e}^{(a_{el}^{\mathrm{mod}}-r)T}-\mathrm{e}^{(a_{el}^{\mathrm{mod}}-r)t_2}\right]\\
&+\frac{E_{el,A}^{\,\mathrm{mod}}+Q_A^{\,\mathrm{mod}}}{\eta_{CHP}^{\mathrm{mod}}}(1+x_{sw,m,was})\mathrm{e}_{coal}^{\mathrm{mod},t=t_2}\frac{1}{a_{coal}^{\mathrm{mod}}-r}\left[\mathrm{e}^{(a_{coal}^{\mathrm{mod}}-r)T}-\mathrm{e}^{(a_{coal}^{\mathrm{mod}}-r)t_2}\right]\\
&+\frac{E_{el,A}^{\,\mathrm{mod}}+Q_A^{\,\mathrm{mod}}}{\eta_{CHP}^{\mathrm{mod}}}\rho_{\mathrm{CO_2}}\,p_{\mathrm{CO_2}}^{\mathrm{mod},t=t_2}\frac{1}{a_{\mathrm{CO_2}}^{\mathrm{mod}}-r}\left[\mathrm{e}^{\left(a_{\mathrm{CO_2}}^{\mathrm{mod}}-r\right)T}-\mathrm{e}^{\left(a_{\mathrm{CO_2}}^{\mathrm{mod}}-r\right)t_2}\right]\\
&+\frac{E_{el,A}^{\,\mathrm{mod}}+Q_A^{\,\mathrm{mod}}}{\eta_{CHP}^{\mathrm{mod}}}\rho_{\mathrm{CO}}\,p_{\mathrm{CO}}^{\mathrm{mod},t=t_2}\frac{1}{a_{\mathrm{CO}}^{\mathrm{mod}}-r}\left[\mathrm{e}^{(a_{\mathrm{CO}}^{\mathrm{mod}}-r)T}-\mathrm{e}^{(a_{\mathrm{CO}}^{\mathrm{mod}}-r)t_2}\right]\\
&+\frac{E_{el,A}^{\,\mathrm{mod}}+Q_A^{\,\mathrm{mod}}}{\eta_{CHP}^{\mathrm{mod}}}\rho_{\mathrm{NO}_X}\,p_{\mathrm{NO}_X}^{\mathrm{mod},t=t_2}\frac{1}{a_{\mathrm{NO}_X}^{\mathrm{mod}}-r}\left[\mathrm{e}^{\left(a_{\mathrm{NO}_X}^{\mathrm{mod}}-r\right)T}-\mathrm{e}^{\left(a_{\mathrm{NO}_X}^{\mathrm{mod}}-r\right)t_2}\right]\\
&+\frac{E_{el,A}^{\,\mathrm{mod}}+Q_A^{\,\mathrm{mod}}}{\eta_{CHP}^{\mathrm{mod}}}\rho_{\mathrm{SO_2}}\,p_{\mathrm{SO_2}}^{\mathrm{mod},t=t_2}\frac{1}{a_{\mathrm{SO_2}}^{\mathrm{mod}}-r}\left[\mathrm{e}^{\left(a_{\mathrm{SO_2}}^{\mathrm{mod}}-r\right)T}-\mathrm{e}^{\left(a_{\mathrm{SO_2}}^{\mathrm{mod}}-r\right)t_2}\right]\\
&+\frac{E_{el,A}^{\,\mathrm{mod}}+Q_A^{\,\mathrm{mod}}}{\eta_{CHP}^{\mathrm{mod}}}\rho_{dust}\,p_{dust}^{\mathrm{mod},t=t_2}\frac{1}{a_{dust}^{\mathrm{mod}}-r}\left[\mathrm{e}^{(a_{dust}^{\mathrm{mod}}-r)T}-\mathrm{e}^{(a_{dust}^{\mathrm{mod}}-r)t_2}\right]\\
&+\frac{E_{el,A}^{\,\mathrm{mod}}+Q_A^{\,\mathrm{mod}}}{\eta_{CHP}^{\mathrm{mod}}}\left(1-u^{\,\mathrm{mod}}\right)\rho_{\mathrm{CO_2}}\mathrm{e}_{\mathrm{CO_2}}^{\mathrm{mod},t=t_2}\frac{1}{b_{\mathrm{CO_2}}^{\mathrm{mod}}-r}\left[\mathrm{e}^{\left(b_{\mathrm{CO_2}}^{\mathrm{mod}}-r\right)T}-\mathrm{e}^{\left(b_{\mathrm{CO_2}}^{\mathrm{mod}}-r\right)t_2}\right]\\
&+\frac{(J+J_M)(1+x_{sal,t,ins})\delta_{serv}^{\,\mathrm{mod}}}{r}\left(\mathrm{e}^{-rt_2}-\mathrm{e}^{-rT}\right)+J_0\left[\left(1+\frac{1}{T}-\frac{t_2}{T}\right)\mathrm{e}^{-rt_2}-\frac{1}{T}\mathrm{e}^{-rT}\right]\\
&+J_M\left[\left(1+\frac{1}{T-t_1}-\frac{t_2}{(T-t_1)}\right)\mathrm{e}^{-rt_2}-\left(1+\frac{1}{T-t_1}-\frac{T}{(T-t_1)}\right)\mathrm{e}^{-rT}\right]\Bigg\}\times\\
&\times\frac{r}{Q_A(1-\mathrm{e}^{-rt_1}])+Q_A^{M}(\mathrm{e}^{-rt_1}-\mathrm{e}^{-rt_2})+Q_A^{\,\mathrm{mod}}(\mathrm{e}^{-rt_2}-\mathrm{e}^{-rT})}\to\min.
\end{aligned}
\tag{5.3}
$$

where:

a_{el}, a_{fuel}, a_{CO_2}, a_{CO}, a_{SO_2}, a_{NO_X}, a_{dust}, b_{CO_2}—exponents showing changes in the prices of electricity, fuel, tariff environmental fees, purchase of CO_2 emission allowances,
$E_{el,A}$—annual gross electricity production,
$e_{el}^{t=0}$, $e_{fuel}^{t=0}$, $e_{CO_2}^{t=0}$, $p_{CO_2}^{t=0}$ etc.—initial values of electricity, fuel, purchase of CO_2 emission allowances, tariff environmental fees,
J—investment expenditure incurred for the construction of the block,
J_M—investment expenditure incurred for the modernization of the block,
Q_A—annual net heat production,
r—interest rate of investment capital (r = 7% was assumed in the calculations),
t_1—time (in years) of the modernization of the block,
t_2—time (in years) of completion of the modernization of the block,
T—expressed in the years, the calculation period of the block's operation (T = 20 years was assumed in the calculations),
u—share of chemical fuel energy in its total annual consumption, for which it is not required to purchase CO_2 emission permits,
$x_{sw,m,was}$—coefficient taking into account the costs of supplementary water, auxiliary materials, sewage disposal, slag storage, waste disposal,
$x_{sal,t,ins}$—factor including costs of wages, taxes, insurance, etc.,
δ_{serv}—annual rate of fixed costs dependent on capital expenditures (maintenance costs, overhauls of equipment),
ε_{el}—indicator of electrical own needs of the block,
η_{CHP}—gross efficiency of the block,
ρ_{CO_2}, ρ_{CO}, ρ_{NO_x}, ρ_{SO_2}, ρ_{dust}—CO_2, CO, NO_x, SO_2, and dust emissions per unit of chemical fuel energy.

Revenues from the sale of electricity produced in the CHP plant with a minus sign $-E_{el,A}(1-\varepsilon_{el})e_{el}^{t=0}\frac{1}{a_{el}-r}\left[e^{(a_{el}-r)t_1}-1\right]$, $-E_{el,A}^{M}(1-\varepsilon_{el}^{M})e_{el}^{M,t=t_1}\frac{1}{a_{el}^{M}-r}\left[e^{(a_{el}^{M}-r)t_2}-e^{(a_{el}^{M}-r)t_1}\right]$, $-E_{el,A}^{\mathrm{mod}}(1-\varepsilon_{el}^{\mathrm{mod}})e_{el}^{\mathrm{mod},t=t_2}\frac{1}{a_{el}^{\mathrm{mod}}-r}\left[e^{(a_{el}^{\mathrm{mod}}-r)T}-e^{(a_{el}^{\mathrm{mod}}-r)t_2}\right]$ constitute the cost of avoided heat production in the combined heat and power plant.

Most often, due to the contracts concluded, the amount of heat supplied to the recipients is constant, and thus $Q_A = Q_A^M = Q_A^{\mathrm{mod}}$. As a result of the modernization of the combined heat and power plant by its repowering with a gas turbine, even in a situation where $Q_A = \text{const}$, the electricity production increases. Its increase depends on the scope and method of modernization. In the case of modernization to the dual-fuel gas-steam system, the sale of electricity from the combined heat and power plant may increase almost twice.

The average unit cost of electricity production $k_{el,av}$ in the modernized unit is also determined by the formula (5.2) from the condition $NPV = 0$, substituting zero for heat production, $Q_R = 0$:

$$
\begin{aligned}
k_{el,av} = \Bigg\{ & \frac{N_{el} t_A}{\eta_{el}} (1 + x_{sw,m,was}) \frac{e_{fuel}^{t=0}}{a_{fuel} - r} \left[e^{(a_{fuel} - r)t_1} - 1 \right] \\
& + \frac{N_{el} t_A}{\eta_{el}} \frac{\rho_{CO_2} p_{CO_2}^{t=0}}{a_{CO_2} - r} \left[e^{(a_{CO_2} - r)t_1} - 1 \right] + \frac{N_{el} t_A}{\eta_{el}} \frac{\rho_{CO} p_{CO}^{t=0}}{a_{CO} - r} \left[e^{(a_{CO} - r)t_1} - 1 \right] \\
& + \frac{N_{el} t_A}{\eta_{el}} \frac{\rho_{NO_X} p_{NO_X}^{t=0}}{a_{NO_X} - r} \left[e^{(a_{NO_X} - r)t_1} - 1 \right] + \frac{N_{el} t_A}{\eta_{el}} \frac{\rho_{SO_2} p_{SO_2}^{t=0}}{a_{SO_2} - r} \left[e^{(a_{SO_2} - r)t_1} - 1 \right] \\
& + \frac{N_{el} t_A}{\eta_{el}} \frac{\rho_{dust} p_{dust}^{t=0}}{a_{dust} - r} \left[e^{(a_{dust} - r)t_1} - 1 \right] + \frac{N_{el} t_A}{\eta_{el}} (1 - u) \frac{\rho_{CO_2} e_{CO_2}^{t=0}}{b_{CO_2} - r} \left[e^{(b_{CO_2} - r)t_1} - 1 \right] \\
& + J(1 + x_{sal,t,ins}) \frac{\delta_{serv}}{r} \left(1 - e^{-rt_1}\right) + zJ \left[1 + \frac{1}{T} - \left(1 + \frac{1}{T} - \frac{t_1}{T}\right) e^{-rt_1} \right] \\
& + \frac{N_{el}^M t_A^M}{\eta_{el}^M} (1 + x_{sw,m,was}) \frac{e_{fuel}^{M,t=t_1}}{a_{fuel}^M - r} \left[e^{\left(a_{fuel}^M - r\right)t_2} - e^{\left(a_{fuel}^M - r\right)t_1} \right] \\
& + \frac{N_{el}^M t_A^M}{\eta_{el}^M} \frac{\rho_{CO_2} p_{CO_2}^{M,t=t_1}}{a_{CO_2}^M - r} \left[e^{\left(a_{CO_2}^M - r\right)t_2} - e^{\left(a_{CO_2}^M - r\right)t_1} \right] \\
& + \frac{N_{el}^M t_A^M}{\eta_{el}^M} \frac{\rho_{CO} p_{CO}^{M,t=t_1}}{a_{CO}^M - r} \left[e^{(a_{CO}^M - r)t_2} - e^{(a_{CO}^M - r)t_1} \right] \\
& + \frac{N_{el}^M t_A^M}{\eta_{el}^M} \frac{\rho_{NO_X} p_{NO_X}^{M,t=t_1}}{a_{NO_X}^M - r} \left[e^{\left(a_{NO_X}^M - r\right)t_2} - e^{\left(a_{NO_X}^M - r\right)t_1} \right] \\
& + \frac{N_{el}^M t_A^M}{\eta_{el}^M} \frac{\rho_{SO_2} p_{SO_2}^{M,t=t_1}}{a_{SO_2}^M - r} \left[e^{\left(a_{SO_2}^M - r\right)t_2} - e^{\left(a_{SO_2}^M - r\right)t_1} \right] \\
& + \frac{N_{el}^M t_A^M}{\eta_{el}^M} \frac{\rho_{dust} p_{dust}^{M,t=t_1}}{a_{dust}^M - r} \left[e^{(a_{dust}^M - r)t_2} - e^{(a_{dust}^M - r)t_1} \right] \\
& + \frac{N_{el}^M t_A^M}{\eta_{el}^M} \left(1 - u^M\right) \frac{\rho_{CO_2} e_{CO_2}^{M,t=t_1}}{b_{CO_2}^M - r} \left[e^{\left(b_{CO_2}^M - r\right)t_2} - e^{\left(b_{CO_2}^M - r\right)t_1} \right] \\
& + \frac{(J + J_M)\delta_{serv}^M}{r} \left(e^{-rt_1} - e^{-rt_2}\right) + zJ \left[\left(1 + \frac{1}{T} - \frac{t_1}{T}\right) e^{-rt_1} - \left(1 + \frac{1}{T} - \frac{t_2}{T}\right) e^{-rt_2} \right] \\
& + J_M \left[\left(1 + \frac{1}{T - t_1} - \frac{t_1}{T - t_1}\right) e^{-rt_1} - \left(1 + \frac{1}{T - t_1} - \frac{t_2}{T - t_1}\right) e^{-rt_2} \right]
\end{aligned}
$$

$$
\begin{aligned}
&+\frac{N_{el}^{\text{mod}}t_{A}^{\text{mod}}}{\eta_{el}^{\text{mod}}}\left(1+x_{sw,m,was}-x_{ch\_fuel\_con}\right)\frac{e_{fuel}^{\text{mod},t=t_2}}{a_{fuel}^{\text{mod}}-r}\left[e^{\left(a_{fuel}^{\text{mod}}-r\right)T}-e^{\left(a_{fuel}^{\text{mod}}-r\right)t_2}\right]\\
&+\frac{N_{el}^{\text{mod}}t_{A}^{\text{mod}}}{\eta_{el}^{\text{mod}}}\left(1-x_{ch\_fuel\_con}\right)\frac{\rho_{CO_2}p_{CO_2}^{\text{mod},t=t_2}}{a_{CO_2}^{\text{mod}}-r}\left[e^{\left(a_{CO_2}^{\text{mod}}-r\right)T}-e^{\left(a_{CO_2}^{\text{mod}}-r\right)t_2}\right]\\
&+\frac{N_{el}^{\text{mod}}t_{A}^{\text{mod}}}{\eta_{el}^{\text{mod}}}\left(1-x_{ch\_fuel\_con}\right)\frac{\rho_{CO}p_{CO}^{\text{mod},t=t_2}}{a_{CO}^{\text{mod}}-r}\left[e^{\left(a_{CO}^{\text{mod}}-r\right)T}-e^{\left(a_{CO}^{\text{mod}}-r\right)t_2}\right]\\
&+\frac{N_{el}^{\text{mod}}t_{A}^{\text{mod}}}{\eta_{el}^{\text{mod}}}\left(1-x_{ch\_fuel\_con}\right)\frac{\rho_{NO_X}p_{NO_X}^{\text{mod},t=t_2}}{a_{NO_X}^{\text{mod}}-r}\left[e^{\left(a_{NO_X}^{\text{mod}}-r\right)T}-e^{\left(a_{NO_X}^{\text{mod}}-r\right)t_2}\right]\\
&+\frac{N_{el}^{\text{mod}}t_{A}^{\text{mod}}}{\eta_{el}^{\text{mod}}}\left(1-x_{ch\_fuel\_con}\right)\frac{\rho_{SO_2}p_{SO_2}^{\text{mod},t=t_2}}{a_{SO_2}^{\text{mod}}-r}\left[e^{\left(a_{SO_2}^{\text{mod}}-r\right)T}-e^{\left(a_{SO_2}^{\text{mod}}-r\right)t_2}\right]\\
&+\frac{N_{el}^{\text{mod}}t_{A}^{\text{mod}}}{\eta_{el}^{\text{mod}}}\left(1-x_{ch\_fuel\_con}\right)\frac{\rho_{dust}p_{dust}^{\text{mod},t=t_2}}{a_{dust}^{\text{mod}}-r}\left[e^{\left(a_{dust}^{\text{mod}}-r\right)T}-e^{\left(a_{dust}^{\text{mod}}-r\right)t_2}\right]\\
&+\frac{N_{el}^{\text{mod}}t_{A}^{\text{mod}}}{\eta_{el}^{\text{mod}}}\left(1-x_{ch\_fuel\_con}\right)\left(1-u^{\text{mod}}\right)\frac{\rho_{CO_2}e_{CO_2}^{\text{mod},t=t_2}}{b_{CO_2}^{\text{mod}}-r}\left[e^{\left(b_{CO_2}^{\text{mod}}-r\right)T}-e^{\left(b_{CO_2}^{\text{mod}}-r\right)t_2}\right]\\
&+\frac{N_{el}^{GT}t_{A}^{\text{mod}}}{\eta_{el}^{TG}}\left(1+x_{sw,m,was}\right)\frac{e_{gas}^{t=t_2}}{a_{gas}-r}\left[e^{\left(a_{gas}-r\right)T}-e^{\left(a_{gas}-r\right)t_2}\right]\\
&+\frac{N_{el}^{GT}t_{A}^{\text{mod}}}{\eta_{el}^{GT}}\frac{\rho_{CO_2}^{gas}p_{CO_2}^{gas,t=t_2}}{a_{CO_2}^{gas}-r}\left[e^{\left(a_{CO_2}^{gas}-r\right)T}-e^{\left(a_{CO_2}^{gas}-r\right)t_2}\right]\\
&+\frac{N_{el}^{GT}t_{A}^{\text{mod}}}{\eta_{el}^{GT}}\frac{\rho_{CO}^{gas}p_{CO}^{gas,t=t_2}}{a_{CO}^{gas}-r}\left[e^{\left(a_{CO}^{gas}-r\right)T}-e^{\left(a_{CO}^{gas}-r\right)t_2}\right]\\
&+\frac{N_{el}^{GT}t_{A}^{\text{mod}}}{\eta_{el}^{GT}}\frac{\rho_{NO_X}^{gas}p_{NO_X}^{gas,t=t_2}}{a_{NO_X}^{gas}-r}\left[e^{\left(a_{NO_X}^{gas}-r\right)T}-e^{\left(a_{NO_X}^{gas}-r\right)t_2}\right]\\
&+\frac{N_{el}^{GT}t_{A}^{\text{mod}}}{\eta_{el}^{GT}}\frac{\rho_{SO_2}^{gas}p_{SO_2}^{gas,t=t_2}}{a_{SO_2}^{gas}-r}\left[e^{\left(a_{SO_2}^{gas}-r\right)T}-e^{\left(a_{SO_2}^{gas}-r\right)t_2}\right]\\
&+\frac{N_{el}^{GT}t_{A}^{\text{mod}}}{\eta_{el}^{GT}}\frac{\rho_{dust}^{gas}p_{dust}^{gas,t=t_2}}{a_{dust}^{gas}-r}\left[e^{\left(a_{dust}^{gas}-r\right)T}-e^{\left(a_{dust}^{gas}-r\right)t_2}\right]\\
&+\frac{N_{el}^{GT}t_{A}^{\text{mod}}}{\eta_{el}^{GT}}\left(1-u^{\text{mod}}\right)\frac{\rho_{CO_2}^{gas}e_{CO_2}^{\text{mod},t=t_2}}{b_{CO_2}^{\text{mod}}-r}\left[e^{\left(b_{CO_2}^{\text{mod}}-r\right)T}-e^{\left(b_{CO_2}^{\text{mod}}-r\right)t_2}\right]\\
&+\frac{(J+J_M)\left(1+x_{sal,t,ins}\right)\delta_{serv}^{\text{mod}}}{r}\left(e^{-rt_2}-e^{-rT}\right)+zJ\left[\left(1+\frac{1}{T}-\frac{t_2}{T}\right)e^{-rt_2}-\frac{1}{T}e^{-rT}\right]\\
&+J_M\left[\left(1+\frac{1}{T-t_1}-\frac{t_2}{T-t_1}\right)e^{-rt_2}-\left(1+\frac{1}{T-t_1}-\frac{T}{T-t_1}\right)e^{-rT}\right]\Bigg\}\times\\
&\times\frac{r}{N_{el}(1-\varepsilon_{el})\left(1-e^{-rt_1}\right)t_A+N_{el}^{M}\left(1-\varepsilon_{el}^{M}\right)\left(e^{-rt_1}-e^{-rt_2}\right)t_A^{M}+N_{el}^{\text{mod}}\left(1-\varepsilon_{el}^{\text{mod}}\right)\left(e^{-rt_2}-e^{-rT}\right)t_A^{\text{mod}}+N_{el}^{GT}\left(1-\varepsilon_{el}^{GT}\right)\left(e^{-rt_2}-e^{-rT}\right)t_A^{\text{mod}}}
\end{aligned}
\tag{5.4}
$$

In the case of the modernization of the block for the dual-fuel gas-steam unit, the power of the gas turbine set N_{el}^{GT} is obviously included in addition to the power of the N_{el} turbine set, and in the operating costs the annual consumption of gas and the cost of emission of pollution resulting from its combustion [see formula (3.2)].

5.2.1 Discussion and Analysis of Calculation Results

The analysis of the economic viability was subject to modernization of the 120 MW amortized block to the dual-fuel gas-steam system. For the purpose of comparison, the modernization of the block to the block for higher parameters of fresh steam was also analyzed. In the calculations, it was assumed that during the modernization ($t_2 - t_1 = 2$), regardless of its method, the block's power is zero, $N_{el}^{M} = 0$ [formula (5.4)]. The block is therefore turned off for 2 years.

In both modernization options, the u share of chemical fuel energy in its total annual consumption was assumed, for which no CO_2 emission allowances are required equal to zero, u = 0, as from 2020 there will no longer be free allowances. It was also assumed that the zero values of the exponents, a_{fuel}, a_{gas}, b_{CO_2} etc., correspond to current prices of coal, gas, purchase of carbon dioxide emission allowances, etc. For example, in an exponent $e_{fuel}(t) = e_{fuel}^{t=0} e^{a_{fuel}t}$ illustrating the change in price of coal prices, its current value is $e_{fuel}^{t=0} = 11.4$ PLN/GJ = 41 PLN/MWh.

Figures 5.2, 5.3, 5.4, 5.5 and 5.6 present the unit costs of electricity production in the amortized unit with the capacity of 120 MW modernized to the dual-fuel gas-steam unit. The power of the gas turbine set is assumed to be equal $N_{el}^{GT} = 120$ MW. Investment expenditures for modernization were assumed in the amount of $J_M =$ PLN 420 million (unit expenditure is $i_M = 3.5$ million PLN/MW). The unit cost of electricity production in the modernized block for current coal and gas prices (i.e. when $a_{fuel} = 0$ and $a_{\text{gas}} = 0$) is relatively slightly higher than the cost in the modernized 120 MW block to the block for higher parameters of fresh steam, in which it is only burned 3 times cheaper coal—Figs. 5.7, 5.8 and 5.9.

As it results from the analysis of the curves in Fig. 5.3, the lower the price of gas to the price of coal, the more advantageous is the fastest possible modernization of the block for the dual-fuel gas-steam system.

Figures 5.7, 5.8 and 5.9 present the unit costs of electricity production in the amortized 120 MW unit, which has been modernized to the block for higher thermal parameters of the fresh steam. The power of the block after modernization is 130 MW. Two investment outlays for modernization were adopted for the calculation: $J_M = 240$ million PLN and $J_M = 360$ million PLN (unit expenditures amount to $i_M = 2$ million/MW and $i_M = 3$ million/MW), for comparison, unitary expenditure on construction "from grass" blocks for supercritical parameters amount to 6.5 million/MW). Analyzing the waveforms in the figures, it is possible to draw the wrong conclusion on account of the economic calculus that modernization is unprofitable, as the unit cost of generating electricity in the modernized unit is the smaller, the year t_1, Fig. 5.1, the start of modernization is more distant, i.e. when modernization will take place as late as possible, and preferably not at all.

The smallest unit cost is for $t_1 = 15$ years. This is because, in the modernized unit, the capital cost in the annual costs of its operation occur again (installment depreciation J_M with interest on it [2]), and of course there were no capital cost in the amortized block. Thus, the annual cost of producing electricity in it is low, because its production is only the cost of operation. However, it should be remembered that

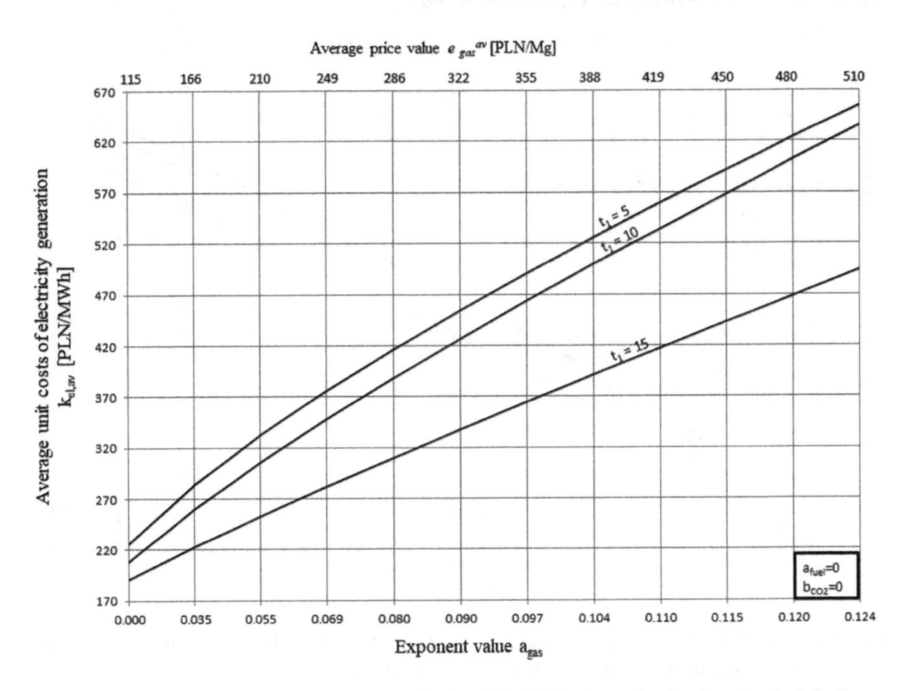

Fig. 5.2 Unit costs of electricity generation in the 120 MW unit modernized to the dual-fuel gas-steam system as a function of the exponent value a_{gas}

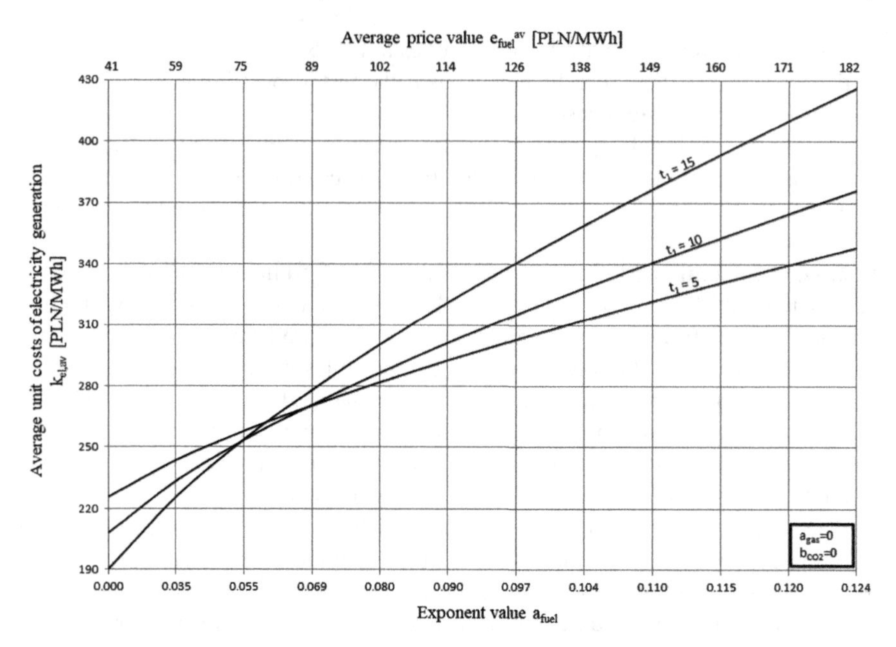

Fig. 5.3 Unit costs of electricity generation in the 120 MW unit modernized to the dual-fuel gas-steam system as a function of the exponent value a_{fuel}

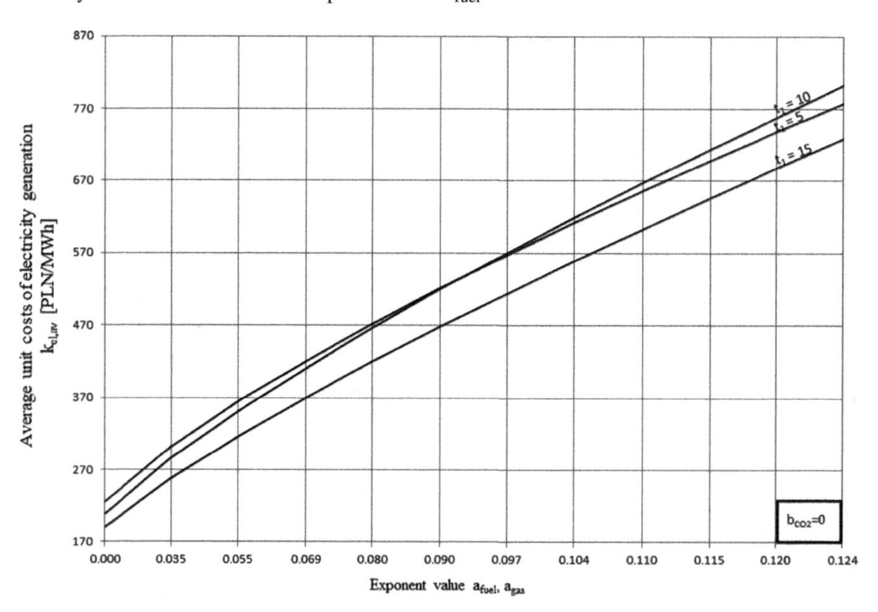

Fig. 5.4 Unit costs of electricity generation in the 120 MW unit modernized to the dual-fuel gas-steam system as a function of the exponents value a_{fuel} and a_{gas}

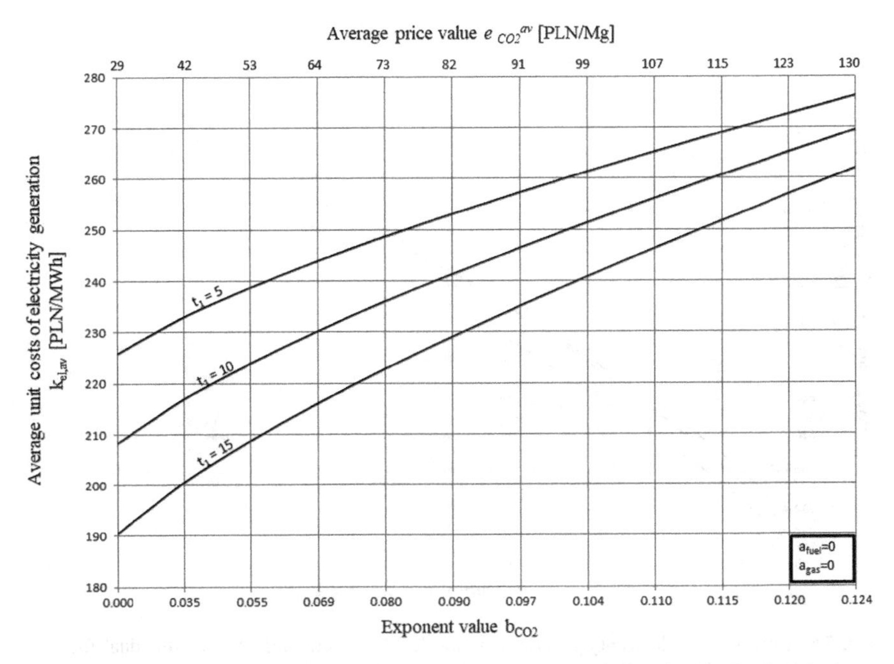

Fig. 5.5 Unit costs of electricity generation in the 120 MW unit modernized to the dual-fuel gas-steam system as a function of the exponent value b_{CO_2}

if the block is not modernized, its further operation will be impossible, it will have to be switched off due to technical wear. Hence the need to modernize as soon as possible.

The results of the conducted analyzes clearly show that an important possibility of modernization of existing coal blocks is their repowering with a gas turbine set despite expensive natural gas. The energy efficiency of such modernized blocks, which is very important, will increase to about 50% and the emission of carbon dioxide per MWh of electricity produced will decrease by half. At the same time, the unit cost of producing electricity in them is also relatively low. For current prices of coal and gas, it is slightly higher than the costs in modernized coal blocks for higher parameters of fresh steam. What's more, and what is very important, such a superstructure doubles the electric power of the block [1, 4]

In addition to the modernization of existing coal-fired units, it is necessary to build "clean", fully economic justified nuclear power plants. A significant increase in the power of the National Power System in Poland is necessary, as the anticipated increase in electricity consumption in Poland is one of the highest in Europe.

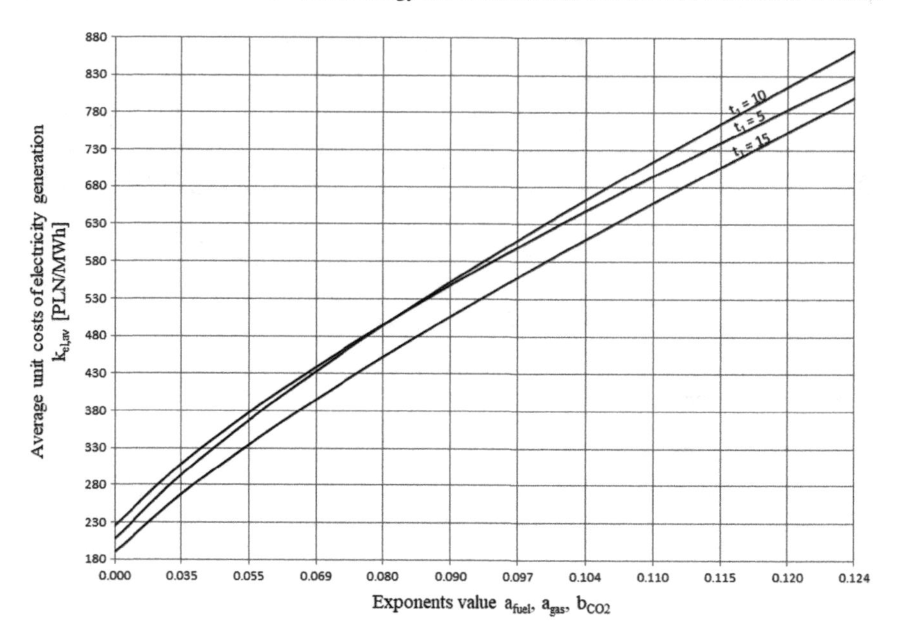

Fig. 5.6 Unit costs of electricity generation in the 120 MW unit modernized to the dual-fuel gas-steam system as a function of the exponents value a_{fuel}, a_{gas} and b_{CO_2}

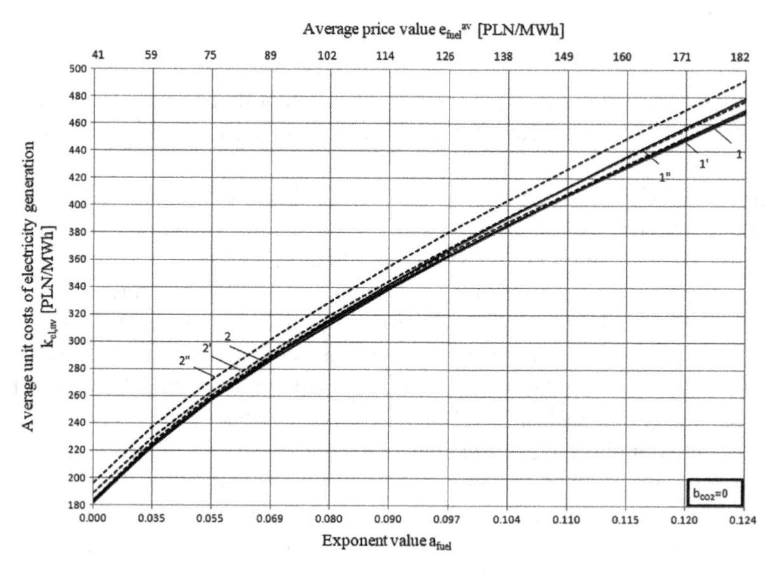

Fig. 5.7 Unit costs of electricity generation in a modernized 120 MW block for a block for higher thermal parameters of fresh steam as a function of exponent a_{fuel} where: 1—applied to $t_l = 15$ years, $i_M = 2$ mln PLN/MW; 1′—applied to $t_l = 10$ years, $i_M = 2$ mln PLN/MW; 1″—applied to $t_l = 5$ years, $i_M = 2$ mln PLN/MW; 2—applied to $t_l = 15$ years, $i_M = 3$ mln PLN/MW; 2′—applied to $t_l = 10$ years, $i_M = 3$ mln PLN/MW; 2″—applied to $t_l = 5$ years, $i_M = 3$ mln PLN/MW

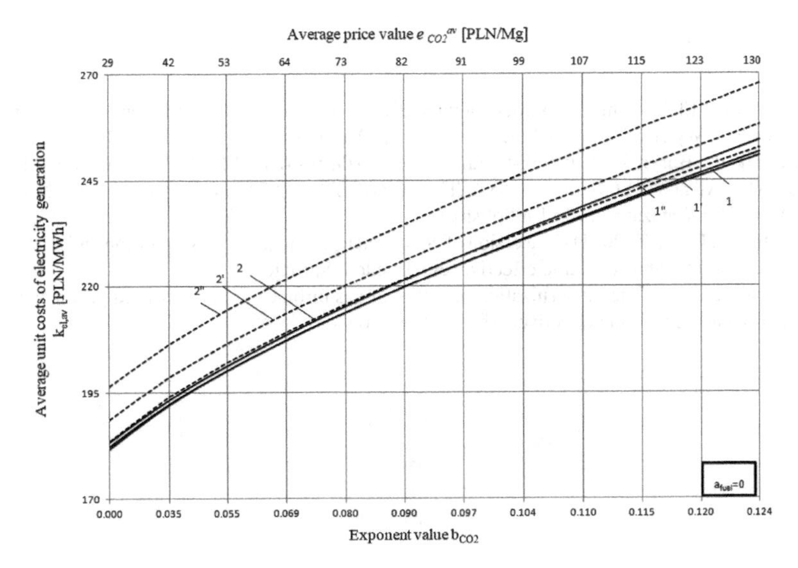

Fig. 5.8 Unit costs of electricity generation in a modernized 120 MW block for a block for higher thermal parameters of fresh steam as a function of exponent b_{CO_2}, where: 1—applied to $t_l = 15$ years, $i_M = 2$ mln PLN/MW; 1′—applied to $t_l = 10$ years, $i_M = 2$ mln PLN/MW; 1″—applied to $t_l = 5$ years, $i_M = 2$ mln PLN/MW; 2—applied to $t_l = 15$ years, $i_M = 3$ mln PLN/MW; 2′—applied to $t_l = 10$ years, $i_M = 3$ mln PLN/MW; 2″—applied to $t_l = 5$ years, $i_M = 3$ mln PLN/MW

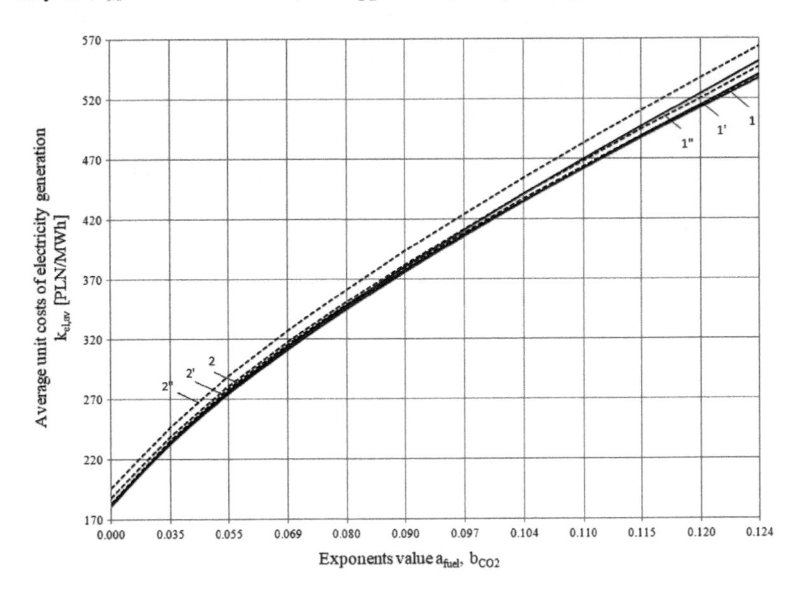

Fig. 5.9 Unit costs of electricity generation in a modernized 120 MW block for a block for higher thermal parameters of fresh steam as a function of exponents a_{fuel} and b_{CO_2}, where: 1—applied to $t_l = 15$ years, $i_M = 2$ mln PLN/MW; 1′—applied to $t_l = 10$ years, $i_M = 2$ mln PLN/MW; 1″—applied to $t_l = 5$ years, $i_M = 2$ mln PLN/MW; 2—applied to $t_l = 15$ years, $i_M = 3$ mln PLN/MW; 2′—applied to $t_l = 10$ years, $i_M = 3$ mln PLN/MW; 2″—applied to $t_l = 5$ years, $i_M = 3$ mln PLN/MW

References

1. Bartnik R.: Elektrownie i elektrociepłownie gazowo-parowe. Efektywność energetyczna i ekonomiczna, Wydawnictwa Naukowo-Techniczne, Warszawa 2009 (dodruk 2012). [in Polish]
2. Bartnik R., Buryn Z., Hnydiuk-Stefan A.: EKONOMIKA ENERGETYKI W MODELACH MATEMATYCZNYCH W ZAPISACH Z CZASEM CIĄGŁYM, Wydawnictwo Naukowe PWN S.A.; Warszawa: 2017 [in Polish]
3. Bartnik R, Buryn Z (2011) Conversion of coal-fired power plants to cogeneration and combined-cycle. Thermal and economic effectiveness. London, Springer
4. Bartnik R (2013) The modernization potential of gas turbines in the coal-fired power industry. Thermal and economic effectiveness. Springer, London

Chapter 6
Impact of the Derogation Mechanism in EU ETS on the Economic Viability of Modernization of Existing Coal Blocks for Dual-Fuel Gas-Steam Systems

The aim of the chapter is to examine the impact on profitability of the so-called derogations, through which have on purpose the possibility of avoiding from the rule of purchase all necessary CO_2 emission allowances at auctions, through the possibility of their free receipt through the implementation of projects affecting the reduction of the level of emissions in the energy sector. Free allowances based on the derogation rule are obtainable upon meeting certain conditions and strictly defined rules, which are described in detail in the further part of the chapter. Currently built, new coal blocks meet the highest standards regarding the emission of pollutants into the air and are less exposed to the cost of purchasing EUA than old, worn out coal blocks. The national energy of each Member State is not able to generate electricity only with the help of newly constructed units, having at its disposal old generation units, which can be successfully modernized using the modernization mechanism defined in art. 10 c of the EU ETS Directive. In the further part of the monograph, the profitability of using this mechanism was examined, consisting in the possibility of obtaining free EUA provided that the modernization of the installation was completed. Power sector in accordance with Directive 2003/87/EC and later 2009/29/EC was covered in 2005 with the CO_2 emission trading system, so-called the EU ETS (European Union Emission Trading Scheme) system. This system includes energy as well as manufacturing companies. The mechanism in the EU ETS system, which is to encourage energy companies to modernize existing sources of electricity and heat, is the mechanism for the allocation of one-time, free carbon dioxide emission allowances by EU officials, so-called derogation mechanism, i.e. one-time reduction of the obligation to pay for each tonne of CO_2 emitted. This voluntary mechanism was introduced in Poland, the Czech Republic, Hungary, Estonia, Lithuania, Latvia, Bulgaria and Romania.

Of course, to obtain such "free tons" of CO_2 (called in the professional environment dealing with the issue of emission trading "EUA"—European Union Allowance) should invest significant funds in the modernization of electricity and heat sources to reduce their emissions CO_2 (due to the millions of funds that have

A. Hnydiuk-Stefan, *Dual-Fuel Gas-Steam Power Block Analysis*, Power Systems,
https://doi.org/10.1007/978-3-030-03050-6_6

to be invested, the words "free tones" are enclosed in quotes). The CO_2 emissions trading market is constantly developing and shaping, under which increasingly strict CO_2 emission limits are being established. In the third trading system period, i.e. in 2013–2020 consisting in particular in the purchase of CO_2 emission permits through government auctions, the abovementioned derogation mechanism was implemented consisting of reduction of the number of tons of CO_2 emissions for which power plant need to buy permits at auctions.

To assess the economic viability of the derogation mechanism, the monograph presents the methodology and results of multi-variant calculations of the unit cost of electricity production in the modernized unit. They were presented with and without consideration of the derogation mechanism. The calculations were carried out for an amortized coal block modernized to an operation with higher parameters of fresh steam with its simultaneous superstructure of a gas turbine and a recovery boiler. Such modernization very significantly affects the reduction of the carbon dioxide emission EF_{CO_2} from the power plant [1–3], even by half (it depends on the power of the gas turbine), which is a necessary condition to use the derogation mechanism. This modernization also increases the power of the block, even more than twice (it is important, as it is necessary to significantly increase the power system's power in Poland) and increases the efficiency of electricity production in it up to 50% [1–4].

In conclusion, the methodology and results of calculations of the unit cost of electricity production in the modernized unit, taking into account and excluding the derogation mechanism, allow to answer the question: does the derogation mechanism under the EU ETS system bring economic benefits to investors, or may have a negligible impact on the modernization profitability?

As already mentioned, the auctioning of allowances is the default method of distributing CO_2 emission allowances in the third period of the CO_2 Emissions Trading Scheme (covering the years 2013–2020) for all installations participating in the European Union Emissions Trading Scheme (EU ETS). However, member states whose electricity systems meet the criteria for the need for retrofitting, in accordance with Article 10 c of the ETS Directive, may temporarily allocate free emission allowances to installations producing electricity, thereby excluding the auctioning of those allowances. Therefore, article 10 c par. 1 of the ETS Directive provides for the possibility to allocate free emission allowances after 2012, until 2019, in connection with the production of electricity only for installations that either operated before December 31, 2008, or whose investment process was actually initiated before that date.

In addition, the installations in question (i.e. those that were physically instituted by the end of 2008) can not be new installations in the light of the ETS Directive. Contents of art. 10a states that "no free allocation shall be made in respect of any electricity produced by new installations". The concept of "new installation" in accordance with art. 1 lit. h of the ETS Directive means "any installation carrying out one or more of the activities listed in Annex 1 to the Directive, which has been authorized to emit greenhouse gases for the first time after 30 June 2011."

The research was undertaken due to the very high volatility of CO_2 emission prices on the trading market and the constant price level of settlement of the derogation mechanism, which may cause various levels of investment profitability. Detailed calculations performed for the needs of that monograph presents the conditions under which an investment using the derogation mechanism is profitable for the investor and those in which deficiencies of CO_2 emission allowances are more profitable to supplement by purchasing allowances on the market.

6.1 Current State of Knowledge

The CO_2 emission trading system was launched in 2005. The energy sector plays a special role in it, as it is the most emission-related segment of the economy in terms of CO_2 volume covered by the EU ETS. This market is still a relatively young market, which is undergoing continuous development and shaping, under which increasingly strict limits on CO_2 emissions are being established. In the third period of the trading system, consisting in particular in the purchase of EUA through government auctions, new market mechanisms were implemented, including derogations based on the possibility of deviating from the purchase of EUAs at auctions, the profitability of which for energy installations was analyzed in detail in this paper. Emissions Trading Scheme (EU ETS) is formerly regulated by Directive 2003/87/EC, and since 2013 by Directive 2009/29/EC so-called EU ETS Directive. In article 10 c of the mentioned Directive there are special conditions for power sector to receive free allocations. The mechanism is described in details in the paper. Member States that can benefit from free allocations under the exception from EUA purchasing on auctions, under Directive 2009/29/EC are: Poland, Czech Republic, Hungary, Estonia, Lithuania, Latvia, Bulgaria and Romania. In the literature on the subject dealing in the field of the EU ETS, many publications on the impact of the EU ETS on investments [1] can be found, as well as attempts to define the assessment of price volatility on the CO_2 market [2, 5]. The unpredictable price levels of EUAs are an important incentive to encourage companies to invest in low-emission technologies.

Further articles on the issue of the EU ETS market moving themes different mechanisms and market reforms affecting the price of EUA [4, 6–8]. Other publications describe the subject of competitive electricity market in the operating conditions of the installations covered by the EU ETS, as since the establishment of the European Union Emission Trading Scheme (EU ETS), carbon prices represent a major cost for EU electricity producers [9]. In [3] the EU ETS is examined in terms of evaluating the EU ETS impacts on profits, investments and prices of the Italian electricity market. In the article [10] initial empirical evidence for the question of investments necessity was made by analyzing corporate reactions to the EU ETS, about whether and in which way the EU ETS affects technology investment decisions that reduce CO_2 emissions. However the analyses concerned whether or not the EU ETS affects on investments and which type of technology companies choose to invest in without taking into account the derogation process mechanism described in article 10 c of the EU ETS Directive.

The issue discussed in this article has not been analyzed so far in publications, but it is extremely important from the investor's point of view and the assessment of the legitimacy of introducing this type of market mechanisms within the EU ETS, whose task is to induce entrepreneurs to take low-emission investments in the energy sector. This article discusses the issue of the profitability of investments within the framework of the available modernization mechanism, taking into account the established way of settling and subsidizing investments converted into free-of-charge EUAs.

6.2 Description of the Modernization Mechanism as a Derogation from the Purchase of the EUAs at Auctions

After 2012 there is no free allocation of allowances in respect of any form of electricity generation, except for cases covered by art. 10 c of the EU ETS Directive 2009/29/EC, moreover, electricity generators can benefit from the provisions of the Free Allowances for electricity generators provided that one of the following conditions is met:

(a) in 2007, the national electricity network was not directly or indirectly connected to the network interconnected system operated by the Union for the Coordination of Transmission of Electricity (UCTE);
(b) in 2007, the national electricity network was only directly or indirectly connected to the network operated by UCTE through a single line with a capacity of less than 400 MW; or
(c) in 2006, more than 30% of electricity was produced from a single fossil fuel, and the GDP per capita at market price did not exceed 50% of the average GDP per capita at market price of the Community.

The condition of qualifying and using the allocation of free CO_2 emission allowances in the field of electricity generation is the fact that the installation operates at the latest on December 31, 2008. In applications submitted in accordance with art. 10 c par. 5 of the Directive, Member States should have demonstrated that installations in their territory considered eligible for the transitional allocation of free emission allowances under Art. 10 c of the Directive fulfill this condition by showing the verified emissions of these installations in the period 2008–2010, at the same time providing the permit number and the account holder of the installation in accordance with the entry to the CITL (Community Independent Transaction Log).

This information should also prove that the installation is still operational. According to art. 10 c par. 3 of the Directive, allocations should be based either on verified emissions from 2005–2007 or on the ex ante efficiency ratio. All eligible installations for which verified emissions data exist for the years 2005–2007 shall be subject to the same allocation methodology. In other words, installations for which there are verified emissions data for the years 2005–2007 should be allocated allowances based on

Table 6.1 Maximum share of free allowances for CO_2 emissions in relation to the years 2005–2007 possible for allocation to electricity producers

Year	Percentage (%)
2013	70
2014	65
2015	60
2016	54
2017	47
2018	39
2019	29
2020	0

Source Own calculations based on EU ETS Directive

these verified emissions or on the basis of the ex ante efficiency ratio. In the absence of such data, the allocation may be based only on the ex ante performance benchmark, however, only one allocation method should be used for each installation. The temporarily allocated free allowances are deducted from the number of allowances that a Member State could in return sell at auction, in accordance with art. 10 par. 2 of the Directive. In 2013, the total number of temporarily allocated free allowances did not exceed 70% of the average annual number of verified emissions for the years 2005–2007 for electricity producers, in relation to the size corresponding to the final national gross consumption of a given Member State. In the following years, the ratio of free allocated allowances is gradually reduced until the total elimination of free allowances in 2020. The Table 6.1 lists the maximum amount of free allowances for CO_2 emissions, separable for installations in individual Member States, against verified emissions for years 2005–2007.

For Member States that did not participate in the Community scheme in 2005, the relevant emissions are calculated on the basis of their verified emissions under the Community scheme in 2007. In addition, Directive 2003/87/EC clearly stated that starting from 2013, auctioning of all emission allowances should become the rule for the energy sector, taking into account its possibilities to charge consumers with alternative costs of CO_2 emissions, within prices electricity, which allows to generate additional income ("extraordinary profits"). Auction sale is aimed at eliminating these extraordinary profits. Article 10 c of the ETS Directive 2009/29/EC contains provisions defining a derogation from a number of important principles of Directive 2003/87/EC, in particular the EU's fully harmonized approach to allocating allowances, introducing auctioning as a basic method for allocating allowances and explicitly excluding the allocation of free emission allowances in with regard to electricity generation. The above rules and regulations are aimed at ensuring the greatest efficiency of the system in economic terms. Article 10 c of the ETS Directive is an exception to the most important principles of Directive 2003/87/EC. In addition, the interpretation and application of this exception should take place in a way that does not undermine the overall objectives of the ETS Directive.

6.2.1 *The Condition to Receive Free EUAs as a Derogation from the Purchase on the Auction*

A condition for granting free emission allowances to installations specified in the Directive is the demonstration of incurring expenditures for investment projects reported in the National Investment Plan (NIP) in the scope of modernization and retrofitting of infrastructure, use of clean technologies, diversification of energy structure or diversification of supply sources. The rules for the implementation of the National Investment Plans include the definition of investments in the national plan that directly or indirectly (investments in networks and auxiliary services) contribute to the reduction of greenhouse gas emissions in a cost-effective manner; they can not, however, strengthen dominant positions or unduly distort competition as well as trade in the internal market and, if possible, should strengthen competition in the internal market in electricity. Investments identified in the national plan should be complementary to investments that Member States must take to achieve other objectives or to meet legal requirements under European Union law. It should also not be investments required to meet the growing demand for electricity.

In addition, investments specified in the national plan should contribute to the diversification of the power structure and sources of energy supply for electricity generation as well as to reduce the intensity of carbon dioxide emissions. Investments should also be economically viable in the absence of allocation of free emission allowances under Art. 10 c of Directive 2003/87/EC upon completion of the transitional allocation of these allowances, with the exception of specific, predefined new technologies at the demonstration stage and listed in Annex III to the Directive.

6.2.2 *The Principle of Allocation of Free CO_2 Emission Allowances for Installations*

The values used to convert eligible investment costs into the number of emission allowances were determined by the European Commission based on the forecast prices of emission allowances in the third settlement period (Guidelines, appendix VI).

Calculation of the value of free emission allowances (EUA) is made at Reference Prices for Poland:

- 14.78 EUR/EUA-for the years 2013–2014
- 20.38 EUR/ EUA-for the years 2015–2019

These values are linked to the emission allowances and the year for which they were granted. However, they do not entail in any way the expense of investment costs.

6.2.3 The Way of Investment Settlement by Installations

Emission allowances are issued to installations generating electricity in the equivalent of the amount of incurred investment costs qualified in the material and financial report to be balanced with the value of emission allowances, however, not more than the number of emission allowances planned to be allocated for a given year in accordance with Table 6.1.

If the amount of incurred investment costs qualified for balancing with the value of emission allowances certified in the material and financial report is higher than the value of emission allowances planned to be allocated for a given year in accordance with Table 6.1, the surplus of investment costs may be used to balance with the value emission rights in the following years of the accounting period, but no longer than for 4 years and no later than by 28 February 2020.

6.2.4 Investment Risk Related to the Case of the Need to Return EUAs Received in the Third Period of the EU ETS

If the approved compliance indicator is not met, the installation carrying out the modernization process is obliged to return the equivalent of the EUA issued in connection with the implementation of the investment task. It is not possible to assess the degree of actual fault of the investor in the event of failure to achieve the assumed compliance ratio—then the investor returns the equivalent of the entire EUA received free of charge.

Similarly, if the investment task is discontinued, the installation shall return the equivalent of the emission allowances received in connection with the implementation of this task until the day of discontinuation of its implementation. In addition, the total value of public assistance granted in the form of EUA for the implementation of a given investment task along with other public aid and de minimis aid may not exceed 100% of the value of eligible costs. The risk related to the possible reimbursement of previously received EUAs is also related to the conversion of the recoverable amount, which is the equivalent of the reimbursable EUA, taking into account the unit price of the EUA, which corresponds to the average price of the EUA listed on ICE/ECX and EEX exchanges on the secondary spot market on the last day in which the EUA trading was conducted, preceding the day of issuance of EUA to the account of the operator, plus interest calculated using the rate of return specified in accordance with Chapter V of Commission Regulation (EC) No. 794/2004 of April 21, 2004 on enforcement Council Regulation (EC) No. 659/1999 laying down detailed rules for the application of Article 93 of the EC Treaty. Receivables paid in connection with the obligation to return the amount equivalent to EUA, constitute the income of the state budget in which the investment took place. The risk of having to return the received EUAs during the investment should be taken into account by the investor in calculating the profitability of the investment.

6.3 Methodology for Calculating the Profitability of Investments

The paper analyzes the investment profitability of the existing installation as part of the modernization aimed at reducing CO_2 emissions. Various options for carrying out investments have been prepared, with particular emphasis on the use of the derogation mechanism on the purchase of EUA through modernization under Art. 10 c of the ETS Directive. Modernization concerns the reconstruction of a coal-fired power plant for a dual-fuel gas-steam power plant. The modernization of the block with its revitalization with simultaneous repowering by the gas turboset and the recovery boiler for the dual-fuel gas-steam system were analyzed [11, 12]. The technologically and technically rational way of modernization of already existing coal blocks, making them modern and at the same time affecting the reduction of the emission factor, is their conversion to dual-fuel gas and steam systems powered with coal and natural gas. Such a conversion is possible by their integration with modern gas technologies based on natural gas-powered gas turbines. The thermal circuit implemented in them will change at that time. In addition to Clausius-Rankine's current steam turbine cycle, the Joule gas turbine cycle will be implemented in them, which will result in a very significant improvement in their energy efficiency. At the same time, the electric power of such modernized blocks will be increased. The emission of pollutants into the natural environment per unit of electricity generated therein will also be significantly reduced as a result of reduced coal consumption and natural gas combustion.

Therefore, the investor will be able to determine at the planning stage the degree of reduction of the emission ratio in order to final investment settlement, which is a prerequisite for the implementation of the investment task as a derogation from the obligation to purchase the entire EUA at auctions. Figure 6.1 presents a time diagram, which was used to build a mathematical model for the analysis of the economic efficiency of the modernization of the analyzed coal power block.

The average unit cost of electricity production in the modernized unit $k_{el,av}$ is calculated from formula (5.4). In this formula (6.1), if the derogation mechanism is considered, the minus sign to the component (6.1) should be added

$$\frac{J_M}{\mathrm{e}_{\mathrm{CO_2}}^{ref}} \mathrm{e}_{\mathrm{CO_2}}^{\mathrm{mod},\, t=t_2}, \tag{6.1}$$

which this mechanism represents, i.e. the one-off mechanism to reduce the cost of purchasing CO_2 emission allowances, where the factor $J_M/\mathrm{e}_{\mathrm{CO_2}}^{ref}$ is the number

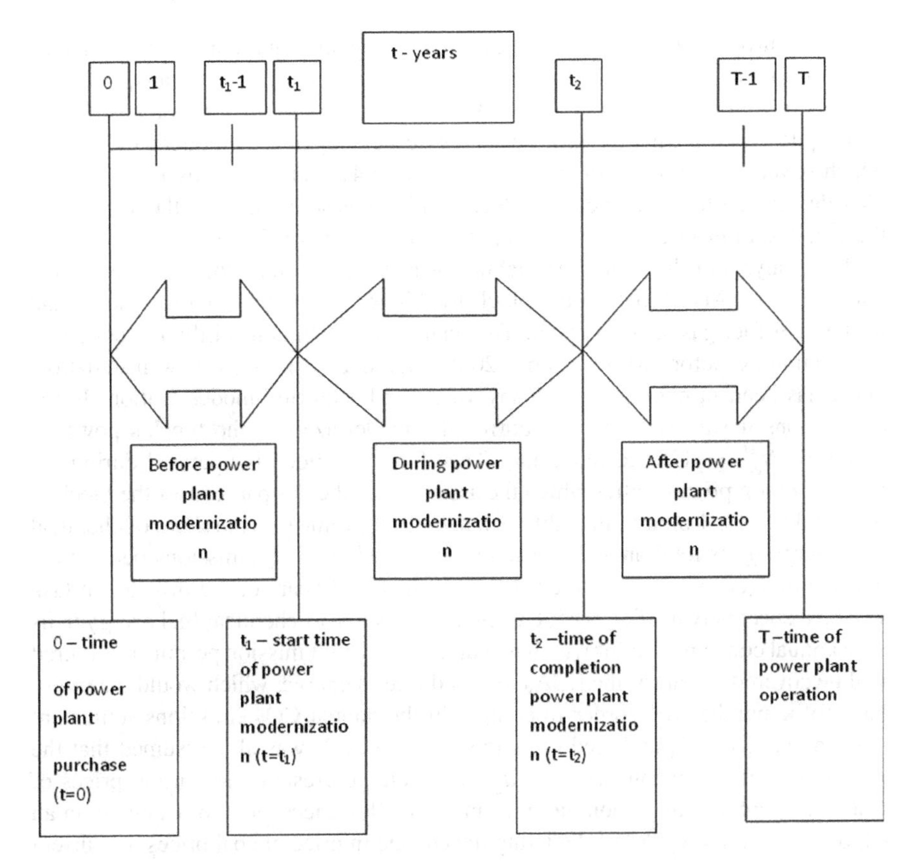

Fig. 6.1 Time diagram of the modernized block's operation

of disposable tonnes of EUA, and the factor e^{-rt_2} is the factor that discounts the component back (6.1) according to the $k_{el,av}$ value calculation rules [13].

As can be seen from formula (6.1), the higher is imposed by the derogation rules the derogation settlement price $e_{CO_2}^{ref}$ (Reference Price) of the carbon dioxide emission, the lower is the amount of free EUA which investor can receive, and thus the value of the derogation mechanism is lower. Currently, the price is $e_{CO_2}^{ref} = 20.38\,€/MgCO_2$ (in the years 2013–2014 the price was much lower and equal to $e_{CO_2}^{ref} = 14.78\,€/MgCO_2$). The value $e_{CO_2}^{\mathrm{mod},\,t=t_2}$ in formula (6.1) means, on the other hand, the purchase price of CO_2 emission allowances at the moment of modernization completion, i.e. the current auction price (primary market) or price from the exchanges, brokers (secondary market). Currently, the price ranges between 5 and 18 €/MgCO_2. If the derogation mechanism is not taken into account in the calculation, component (6.1) in formula (5.4) should be omitted.

In the case of modernization of the block for a dual-fuel gas-steam unit, the power of the gas turbine set must be taken into account in addition to the power of the steam turbine set, and in the operating costs the annual gas consumption and the emission of pollution resulting from its combustion [11, 14]. An important possibility of modernization of existing coal blocks is their superstructure with a gas turbine set despite expensive natural gas. The energy efficiency of such modernized blocks, which is very important, will increase from 32 to 41% and the emission of carbon dioxide per MWh of electricity produced will decrease by half. At the same time, the unit cost of electricity generation in them is also relatively low [15].

The analysis of the economic viability was subject to the modernization of the amortized 120 MW block to the block for higher parameters of fresh steam and to the dual-fuel gas-steam system. The analyzed installation fulfills the condition of functioning before 31 December 2008, moreover it is not a new installation, but it has been in operation for many years, and requiring modernization. In the calculations it was assumed that during the modernization, the block's power is 120 MW, $N_{el}^{M} = 120$ (formula 5.4). Therefore, the block is operated during the modernization process, after which the assembly of the gas part causes the block to be shut down for 1 month. In addition, a decreasing value of u the share of chemical fuel energy in its total annual consumption, for which CO_2 emissions permits are not required, according to Table 6.1 for the analyzed years of the investment task will be respectively $u = 0.7, 0.65, 0.6$, etc., as the share of chemical fuel energy in its total annual consumption for which no purchase of CO_2 emission permits is required will be covered as part of the EUAs received free of charge, which would otherwise have to be purchased in order to comply in the annual CO_2 emissions settlement with the number of purchased emission allowances. It was also assumed that the zero values of the exponents, $a_{\mathrm{fuel}}, a_{\mathrm{gas}}, b_{\mathrm{CO_2}}$ etc., correspond to current prices of coal, gas, purchase of carbon dioxide emission allowances, etc. For example, in an exponent $e_{fuel}(t) = e_{fuel}^{t=0} e^{a_{fuel} t}$ depicting the change in price of coal prices, its current value is $e_{\mathrm{fuel}}^{t=0} = 2.7$ EUR/GJ. The power of the gas turbine set is assumed to be equal $N_{el}^{GT} = 120\,\mathrm{MW}$. Investment expenditures for modernization were assumed in the amount of $J_M = 100$ million euros (unit expenditures amount to $iM = 0.83$ million EUR/MW). In the calculations it was assumed that the power of the block (i.e. steam turbine set) is after modernization $N_{el}^{\mathrm{mod}} = 130\,\mathrm{MW}$, and the power of the gas turbine set is equal to $N_{el}^{GT} = 120\,\mathrm{MW}$ (formula 5.4).

The analyzed case refers to the installation, the operation of which will not be possible without modernization, it will have to be switched off due to technical wear and condition. Hence the need to modernize as soon as possible. In addition, modernization expenditures, even when replacing boilers and steam turbines with new ones with higher thermal parameters of fresh steam, will be small compared to the expenditures for new coal units with supercritical parameters (1.55 million EUR/MW), in which the entire power plant infrastructure must be built, and which already exists in modernized ones. The analyzed cost-effectiveness of the mechanism under the EU ETS applies to existing power plants. It is most advantageous to use it in those installations that can be modernized in the most effective way. At most, it will also need to be revitalized. The unit costs of electricity generation in such modernized blocks, even if the necessary CO_2 purchase permits are then clearly lower than the costs in new coal units. Since this installation would be revitalized even without the use of the EU ETS mechanism, calculations of the specific cost of electricity production in the case of investments realized independently by the investor and without the use of state aid were made, as well as for a few selected investment options using the EU ETS mechanism (Figs. 6.5, 6.6, 6.7, 6.8, 6.9, 6.10, 6.11, 6.12, 6.13, 6.14, 6.15 and 6.16). The number of possible scenarios of changes in fuel prices and tonnes of CO_2 emissions is unlimited, however, the models used allow estimating the value of investments based on various variables.

6.3.1 The Case of Using the Derogation Mechanism from the Purchase of EUA at Auctions

The essence of the analysis presented in this chapter is a comparison of the unit cost of electricity generation, taking into account the use of the derogation mechanism from the obligation to purchase EUAs at auctions. Assuming that the modernization is carried out for a period of 5 years and considering the different price levels of EUA settlements accounted for under the EU ETS mechanism, calculations of investment profitability by two analyzed periods were made for the following data:

(1) 2013–2017 (this period will include settlement of free allocation of EUA at a price of 14.78 € in 2013–2014 and 20.38 € in –2017)
(2) years –2019 (this period will include settlement of free allocation of EUA at a price of 20.38 € in –2017)

In 2020, the installation under consideration will not receive any free EUAs even if there is a surplus of investment costs in previous years, which could be converted into a free allocated EUA in the previous years. The value of return on investment under the EU ETS mechanism depends on the price levels set in the mechanism when settling investments and on the current market value of the EUA, which is the main element of the consideration of investment profitability in this article. An important factor will be the market price of the EUA at which the investor would purchase the EUA at governmental auctions or through the exchange, without using the derogation mechanism. An important incentive for the modernization of installations for power plants covered by the EU ETS is the simultaneous reduction of CO_2 emissions as a result of the investment, which will effect into the costs of further operation of the power block. Other factors such as the cost of investment per unit of power, the price of electricity sales, variable price relations of individual factors affecting the unit cost of electricity generation and relations between them have a significant impact on the investment decision regarding the modernization of the power block and the time it takes to apply mechanism of reimbursement within the EU ETS as well as without it.

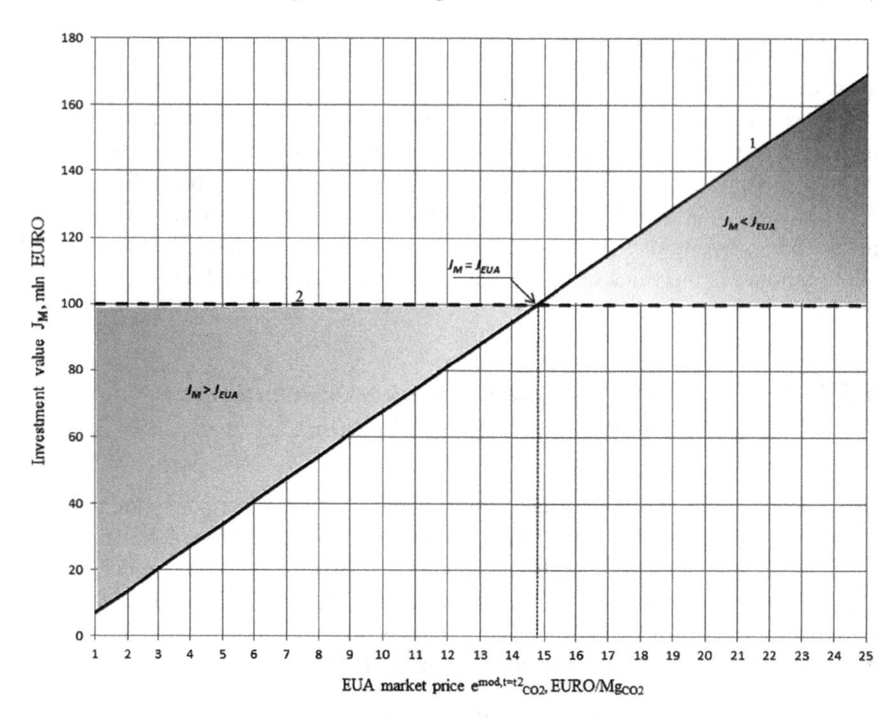

Fig. 6.2 Chart of investment profitability depending on the EUA price, where: 1—real investment value converted into the EUA price from the auction or exchange market $e_{CO_2}^{mod,\,t=t_2}$, 2—real investment value converted into the EUA settlement price as part of the EU ETS derogation mechanism $e_{CO_2}^{mod,\,t=t_2} = e_{CO_2}^{ref\,1}$

6.3.1.1 Profitability of Modernization as Part of the Derogation from Purchase at Auctions in 2013–2014

In the investment period falling in the years 2013–2014, the settlement price (Reference Price $e_{CO_2}^{ref\,1}$) for free allocations is set at 14.78 EUR/EUA. Investment expenditure for the modernization under consideration is J_M = EUR 100 million, this means that the free allocation of EUA after completing the investment project will amount to over 6.766 million EUAs. Figure 6.2 presents a graph of the actual return of investment costs depending on the value of EUA on the secondary market.

As shown in Fig. 6.2, in order for the investment to be profitable for the investor in 2013–2014, the level of EUA purchase on the auction or exchange market $e_{CO_2}^{mod,\,t=t_2}$ should be at least equal to the settlement price under the derogation mechanism $e_{CO_2}^{ref\,1}$. Then the total cost of purchasing EUA (J_{EUA}) will correspond to the reimbursed investment cost (J_M).

The difference in the value of reimbursement of incurred investment outlays as part of the RZ_D derogation can therefore be considered as:

$$RZ_D = J_M * (e_{CO_2}^{mod,\,t=t_2} / e_{CO_2}^{ref\,1}) - J_M \tag{6.2}$$

where:

RZ_D [euro]—the difference in the investment cost resulting from the investment under the derogation, where the negative value means the real lower investment costs covered by derogation mechanism so the investor has to cover the difference, and the positive value will mean an additional profit for the investor through derogation compare to EUA purchase on the market, in the case of zero, the total cost of investment will be covered by the EUAs value received free of charge in derogation.

J_M [euro]—the value of investments in modernization.

$e_{CO_2}^{mod,\,t=t_2}$ [euro/EUA]—the value of EUA on the current primary or secondary market (auction or auction)

$e_{CO_2}^{ref\,1}$—settlement price under the derogation mechanism (14.78 EUR/EUA in years 2013–2014).

Figure 6.3 shows the dependence of the price on the secondary market on the actual value of the subsidy obtained as part of the derogation. The value RZ_D at which the additional profit is 0 means that the investment was entirely covered by the derogation mechanism and the investor did not receive any additional profits. Values RZ_D above 0 mean that the investment cost was completely covered by the derogation mechanism and the investor gained additional profits and at the same time it means that the purchase of EUA on the secondary market would be more expensive by the value presented on the Fig. 6.3. Therefore, the investor would incur the cost of purchasing the EUA corresponding to the value of the investment and the surplus located above the point 0 on the x axis in Fig. 6.3. Negative value means to what extent the investor covered the investment cost from its own funds. For example, with the market price $e_{CO_2}^{mod,\,t=t_2}$ of 3 EUR/EUA, the investor incurred an investment cost of 80 million euros, so only 20 million euros were covered by the derogation, because the amount of EUA that the investor received free of charge is worth only 20 million euros at the moment of receive it.

6.3.1.2 Profitability of Modernization as Part of the Derogation from Purchase at Auctions in 2015–2019

In the investment period falling in the years 2015–2019, the settlement price (Reference Price $e_{CO_2}^{ref\,2}$) for free allocations is set at 20.38 EUR/EUA. Investment expenditure for the modernization under consideration is J_M = EUR 100 million, this means that the free allocation of EUA after completing the investment project will amount to over 4.9 million EUAs. Figure 6.4 presents a graph of the actual return of investment costs depending on the value of EUA on the secondary market.

As shown in Fig. 6.4, for the investment to be profitable for the investor in 2015–2019, the level of EUA purchase on the auction or exchange market $e_{CO_2}^{mod,\,t=t_2}$ should be at least equal to the settlement price under the derogation mechanism

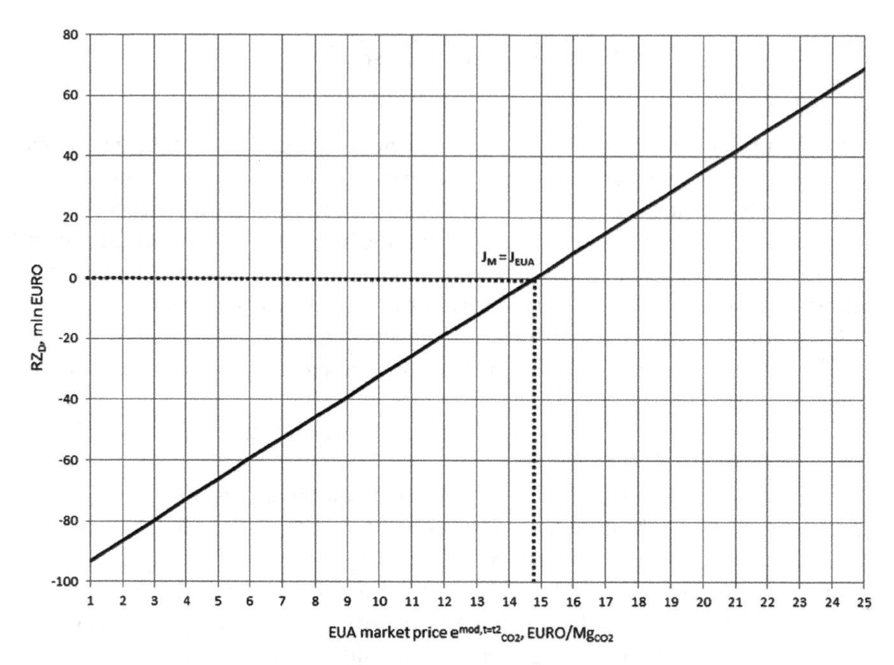

Fig. 6.3 Dependence of the EUA price on the market $e_{CO_2}^{mod,\,t=t_2}$ on the real return on capital value expenditure for modernization as part of the derogation from the purchase at auctions

$e_{CO_2}^{ref2}$. Then the cost of purchasing EUA (J_{EUA}) will correspond to the reimbursed investment cost (J_M).

Similarly to the formula (6.2), the difference in the value of the return of incurred investment outlays as part of the RZ_D derogation can therefore be considered as:

$$RZ_D = J_M * (e_{CO_2}^{mod,\,t=t_2}/e_{CO_2}^{ref2}) - J_M \tag{6.3}$$

where:

RZ_D [euro]—the difference in the investment cost resulting from the investment under the derogation, where the negative value means the real lower investment costs covered by derogation mechanism so the investor has to cover the difference, and the positive value will mean an additional profit for the investor through derogation compare to EUA purchase on the market, in the case of zero, the total cost of investment will be covered by the EUAs value received free of charge in derogation.

J_M [euro]—the value of investments in modernization.

$e_{CO_2}^{mod,\,t=t_2}$ [euro/EUA]– the value of EUA on the current primary or secondary market (auction or auction)

$e_{CO_2}^{ref2}$—settlement price under the derogation mechanism (20.38 euro/EUA in years 2015–2019).

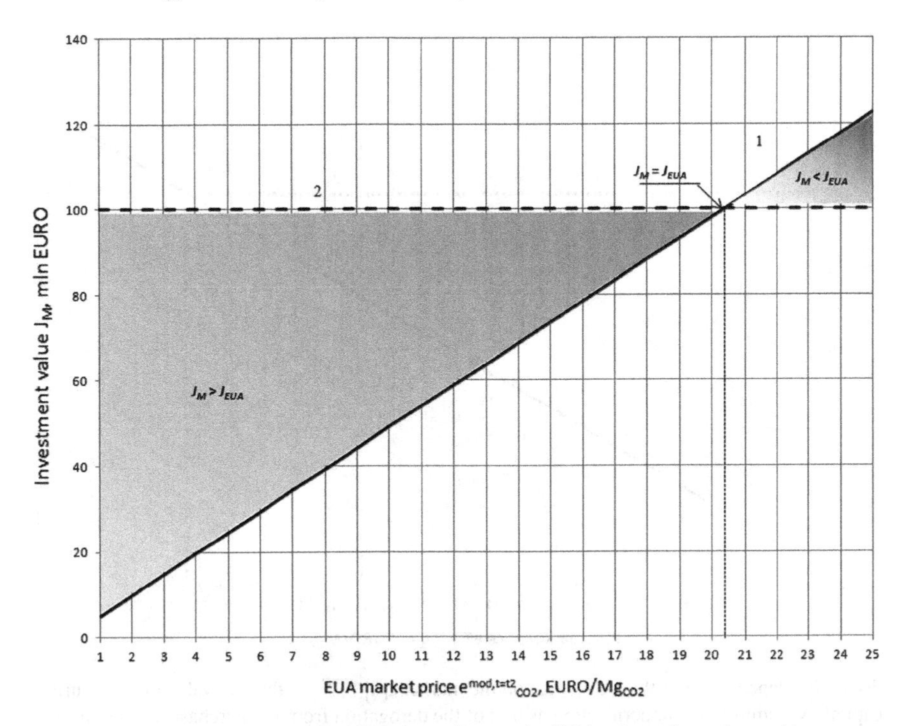

Fig. 6.4 Chart of investment profitability depending on the EUA price, where: 1—real investment value converted into the EUA price from the auction or exchange market $e^{\text{mod},\, t=t_2}_{CO_2}$, 2—real investment value converted into the EUA settlement price as part of the EU ETS derogation mechanism $e^{\text{mod},\, t=t_2}_{CO_2} = e^{ref2}_{CO_2}$

Figure 6.5 shows the dependence of the secondary market price on the actual value of the subsidy obtained as part of the derogation. The value at which the additional profit is 0 means that the investment was entirely covered by the derogation mechanism and the investor did not receive any additional profits. Values above 0 mean that the investment was completely covered by the derogation mechanism and the investor gained additional profits and at the same time it means that the purchase of EUA on the secondary market would be more expensive. Thus, the investor would incur the cost of purchasing the EUA corresponding to the value of the investment and the surplus located above the point 0 on the x axis of the chart (Fig. 6.5). Negative value means to what extent the investor covered the investment cost from its own funds. For example, with the secondary market price of 3 euros, the investor incurred an investment cost of 85 million euros, so only 15 million euros were covered by the derogation, because the amount of EUA that the investor received free of charge is worth only 15 million euros at the time of receipt.

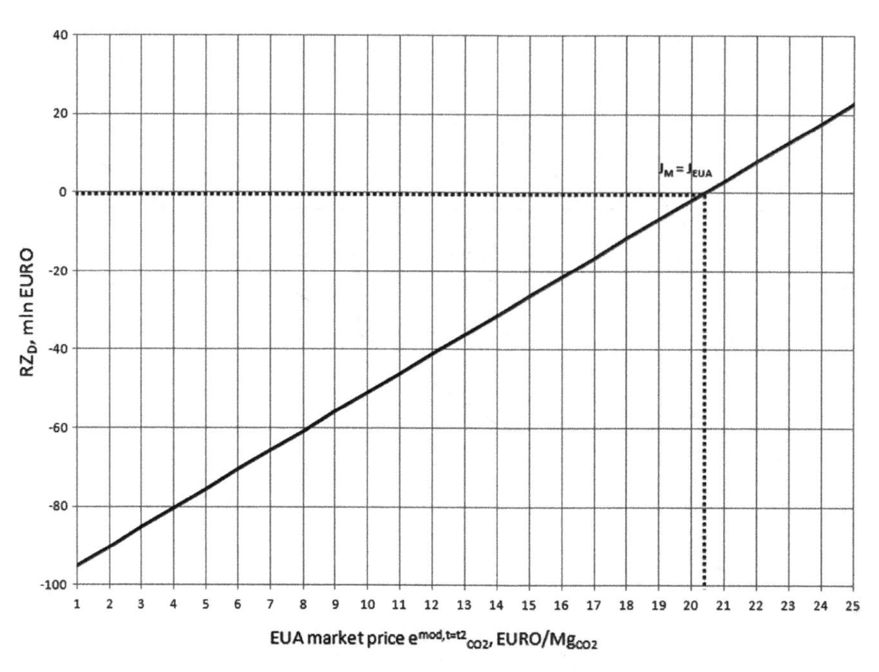

Fig. 6.5 Dependence of the EUA price on the market $e_{CO_2}^{mod,\, t=t_2}$ on the real value of the return on capital expenditure for modernization as part of the derogation from the purchase at auctions

6.4 Discussion and Analysis of Calculation Results

Based on formula 5.4, calculations of the unit cost of electricity generation in various assumed price scenarios were made. The input data for the analyzed installation is summarized in Table 6.2.

The graphs (Figs. 6.6, 6.7, 6.8, 6.9, 6.10, 6.11, 6.12, 6.13, 6.14, 6.15, 6.16 and 6.17) present the results of calculations of the profitability of using the derogation mechanism. EUA market prices $e_{CO_2}^{mod,\, t=t_2}$ corresponding to 5, 10, 15 and 20 euro/EUA were adopted in order to compare the expenditures on the purchase of EUA with the cost of modernizing the installation. In addition, three cases of modernization were analyzed:

- using the derogation mechanism within the first settlement period (covering the years 2013–2014),
- using the derogation mechanism within the second period (covering the years 2015–2019),

without support from the derogation mechanism, which means that the investor decides to cover all costs from own resources.

Calculations can provide the investor with valuable information on the cost-effectiveness of spending under the derogation mechanism instead of the purchase

Table 6.2 Summary of input data for the analyzed power unit

Gross capacity of gas power unit $N_{el} = 120$ MW
Investment for modernization $J_M = 100$ mln EURO
Annual exploitation time of power unit $t_A = 7500$ h
Internal electrical load of coal power unit: $\varepsilon_{el} = 7.6\%$, for gas unit $\varepsilon_{el\_gas} = 4.0\%$
Discount rate $r = 7\%$, rate of income tax $p = 19\%$
Construction period of power unit $b = 5$ years
Exploitation time $T = 20$ years
Annual rate of maintenance and overhaul $\delta_{serv} = 3\%$
Total cost $x_{sal,t,ins} = 0.25$, total cost $x_{sw,m,was} = 0.02$
Specific coal price $e_{fuel} = 2.7$ EURO/GJ, gas price $e_{gas} = 7.6$ EURO/GJ
Price per CO_2 emissions EUA: $e_{CO_2} = 5\,EURO/Mg_{CO_2}$
Tariff rates per emissions of: $p_{CO_2} = 0.07\,EURO/Mg_{CO_2}, p_{CO} = 26 EURO/Mg_{CO}, p_{NO_x} = 126\,EURO/Mg_{NO_X}, p_{SO_2} = 126\,EURO/Mg_{SO_2}, p_{dust} = 83\,EURO/Mg_{dust}$
Emissions from carbon coal combustion: $\rho_{CO_2} = 95\,kg/GJ$, $\rho_{CO} = 0.01\,kg/GJ$, $\rho_{NO_x} = 0.164\,kg/GJ$, $\rho_{SO_2} = 0.056\,kg/GJ$, $\rho_{dust} = 0.007\,kg/GJ$. Emissions from gas combustion: $\rho_{CO_2} = 55\,kg/GJ$, $\rho_{CO} = 0\,kg/GJ$, $\rho_{NO_x} = 0.02\,kg/GJ$, $\rho_{SO_2} = 0\,kg/GJ$, $\rho_{dust} = 0\,kg/GJ$

of EUA on the secondary market. In addition to the level of EUA market prices, an important factor is also the start time of the investment of the modernized installation. In the calculations, a time interval was assumed for the start of modernization of the installation for 5, 10 and 15 years since its launch. The figures also assume various average purchase prices of EUAs during the further operation of the block, between 5 and 22 EUR/EUA. This is an extremely important factor, as a result of the modernization of the installation, the emission factor will be reduced per unit of electricity produced therein as a result of reduced coal consumption and natural gas combustion.

6.4.1 *Impact of Changes in EUA Market Prices on Investment Profitability as Part of the Derogation Mechanism*

While performing calculations for the assessment of the impact of changes in the EUA price during the operation of the unit after modernization, it was assumed that gas prices will remain at a constant level of 7.6 EUR/GJ, while the prices of coal will remain constant at 2.7 EUR/GJ.

At low EUA market prices during the investment, the start time of the investment is important factor affecting profitability. The later investment will start, the better, however, as EUA prices on the market increase further and need to cover their

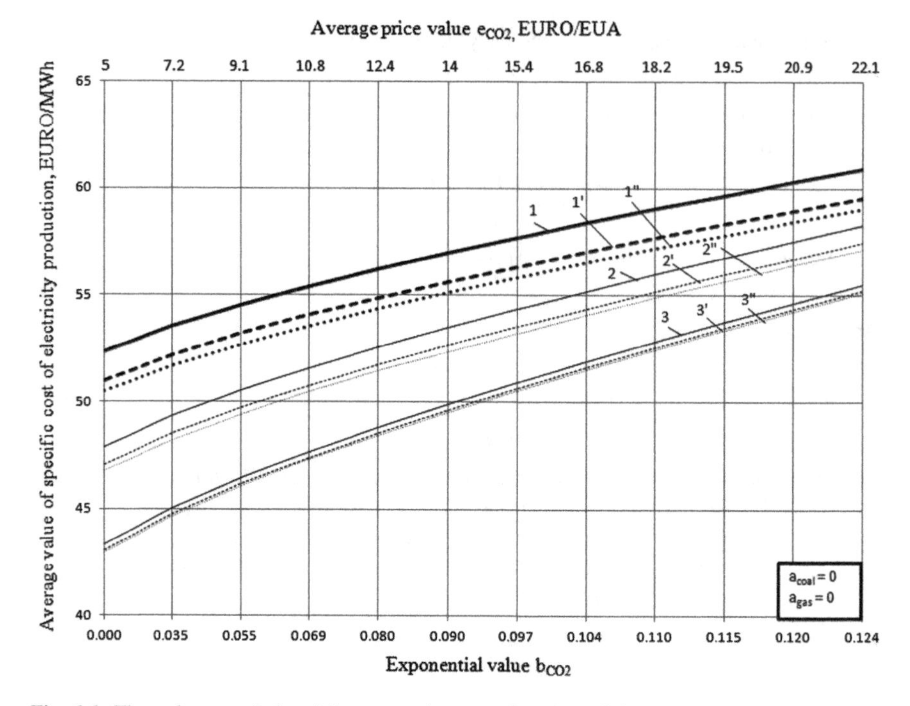

Fig. 6.6 The unit cost of electricity generation as a function of the exponent value of b_{CO_2} at the EUA market price = 5 € during the investment period, where: 1, 2, 3—respectively for t1 = 5 years, t1 = 10 years, t1 = 15 years, without a derogation mechanism; 1′, 2′, 3′—respectively for t1 = 5 years, t1 = 10 years, t1 = 15 years, for the derogation mechanism at the settlement price of the investment 20.38 euro/EUA (period 2015–2019); 1″, 2″, 3″—respectively for t1 = 5 years, t1 = 10 years, t1 = 15 years, for the derogation mechanism at the settlement price of the investment 14.78 euro/EUA (period 2013–2014)

shortage by purchasing on the EU ETS market, the investment in modernization of the installation will be necessary as it will allow the investor to reduce the emission rate during electricity production. The derogation mechanism in this case is moderately favorable for application, especially in the perspective of the implementation of 15 years from the start-up of the installation, because it only slightly affects the reduction of the cost of electricity generation. It slightly affects the reduction of this cost when the short investment start time has elapsed since the installation was launch and for the lower settlement price of the EUAs thus received free of charge.

As can be seen in Fig. 6.7, assuming a market price of 10 EUR/EUA during the investment, it is beneficial to make a decision on modernization later. However, already at the price level of 15 euro/EUA for emission allowances during further operation of the installation, faster modernization will be more beneficial. The most expensive option is, with these pricing assumptions, the implementation of investments from own funds with a fast start time of the investment since the installation was started.

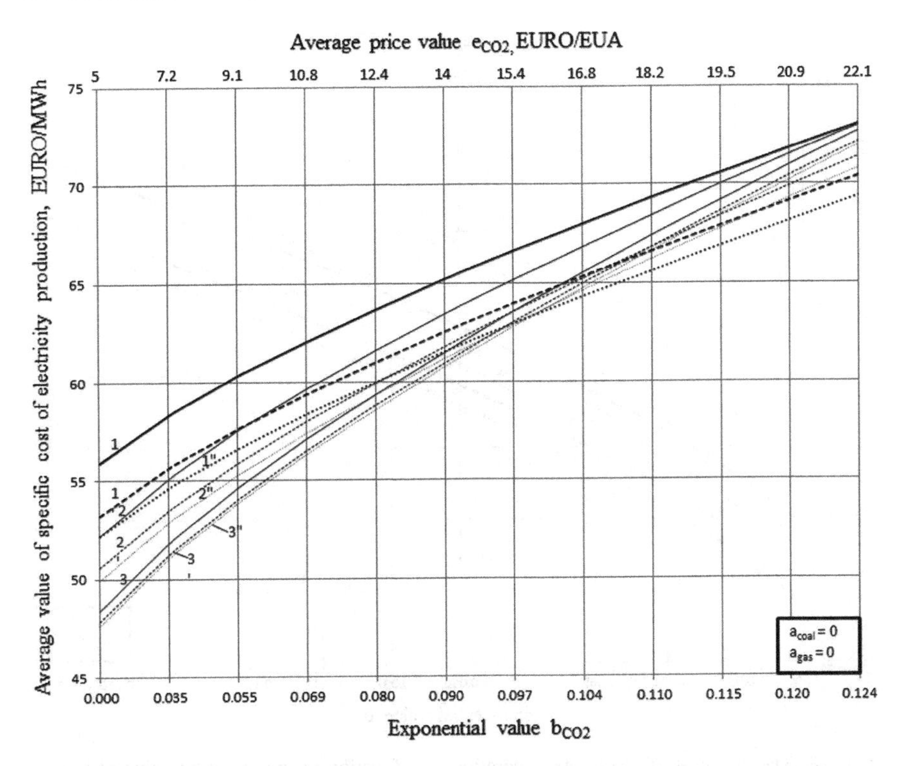

Fig. 6.7 The unit cost of electricity generation as a function of the exponent value of b_{CO_2} at the EUA market price = 10 € during the investment period, where: 1, 2, 3—respectively for t1 = 5 years, t1 = 10 years, t1 = 15 years, without a derogation mechanism; 1′, 2′, 3′—respectively for t1 = 5 years, t1 = 10 years, t1 = 15 years, for the derogation mechanism at the settlement price of the investment 20.38 euro/EUA (period 2015–2019); 1″, 2″, 3″—respectively for t1 = 5 years, t1 = 10 years, t1 = 15 years, for the derogation mechanism at the settlement price of the investment 14.78 euro/EUA (period 2013–2014)

In the case of the EUA price of 15 EUR during the investment and maintenance of the modernized installation at this level or even further increases, the most advantageous option for the investor is the fastest possible modernization using the derogation mechanism from the period when the settlement rate is 14.78 euro/EUA. In the case presented in Fig. 6.8, using the derogation mechanism even after a higher settlement rate of 20.38 EUR/EUA and concurrently realizing the investment within 10 years from the start-up of the installation turns out to be more profitable than using only the investor's resources and faster investment implementation, i.e. within 5 years of the installation's operation.

Figure 6.9 shows that at the EUA price of 20 EUR/EUA during the investment and price levels of the EUA for the further operation of the installation at the level of 5 EUR/EUA using the derogation mechanism, the investment start time is of the utmost importance. The faster the investment will be realized, the more advantageously it

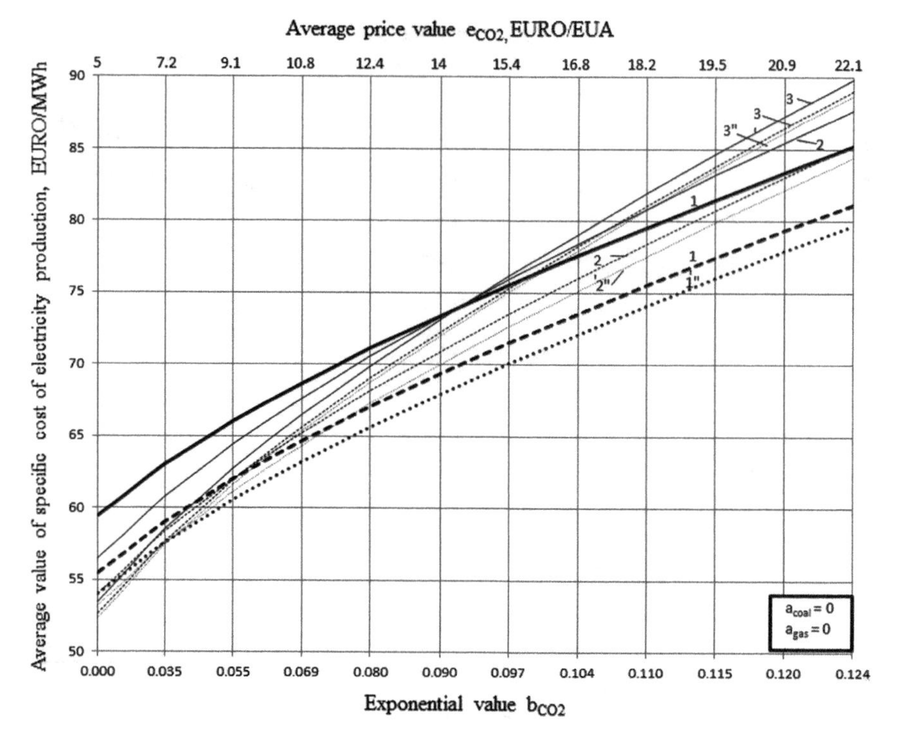

Fig. 6.8 The unit cost of electricity generation as a function of the exponent value of b_{CO_2} at the EUA market price = 15 € during the investment period, where: 1, 2, 3 - respectively for t1 = 5 years, t1 = 10 years, t1 = 15 years, without a derogation mechanism; 1′, 2′, 3′—respectively for t1 = 5 years, t1 = 10 years, t1 = 15 years, for the derogation mechanism at the settlement price of the investment 20.38 euro/EUA (period 2015–2019); 1″, 2″, 3″—respectively for t1 = 5 years, t1 = 10 years, t1 = 15 years, for the derogation mechanism at the settlement price of the investment 14.78 euro/EUA (period 2013–2014)

will be. This dependence is of particular importance for the investment realized only from the investor's own resources at much higher levels of EUA prices exceeding 18 EUR/EUA.

6.4.2 *Impact of Gas Price Changes on the Profitability of the Derogation Mechanism*

Another aspect that was considered in this chapter is the impact of gas price changes on specific cost of electricity production. This is of particular importance as the installation after modernization will use gas as a fuel in addition to coal combustion. From among an unlimited number of possibilities, various price scenarios for the purchase of gas during the operation of the power block after modernization from

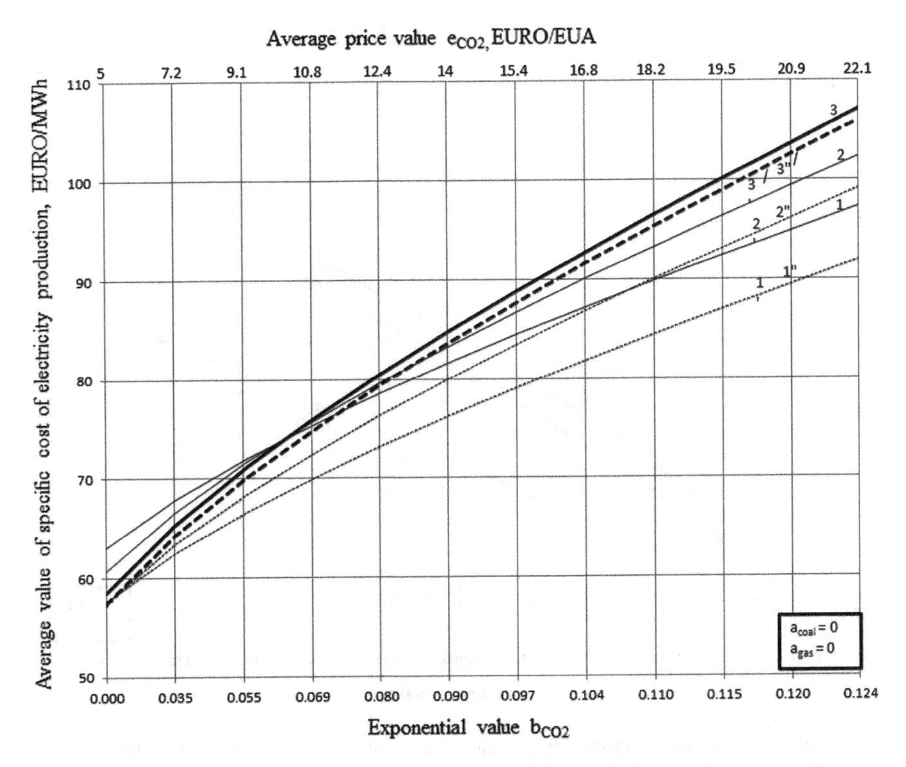

Fig. 6.9 The unit cost of electricity generation as a function of the exponent value of b_{CO_2} at the EUA market price = 20 € during the investment period, where: 1, 2, 3—respectively for t1 = 5 years, t1 = 10 years, t1 = 15 years, without a derogation mechanism; 1′, 2′, 3′—respectively for t1 = 5 years, t1 = 10 years, t1 = 15 years, for the derogation mechanism at the settlement price of the investment 20.38 euro/EUA (period 2015–2019); 1″, 2″, 3″—respectively for t1 = 5 years, t1 = 10 years, t1 = 15 years, for the derogation mechanism at the settlement price of the investment 14.78 euro/EUA (period 2013–2014)

the range of 8–34 EUR/GJ have been assumed. Moreover, it is assumed that during the operation of the modernized coal-fired block, the prices of coal will remain at a constant level of 2.7 EURO/GJ, while the EUA prices will remain stable at 5 EUR/EUA.

As shown in the graphs above (Figs. 6.10 and 6.11), assuming rising gas prices at the same time at low EUA prices for the duration of investments of 5 or 10 euros/EUA and with further EUA levels remaining at constant $b_{CO_2} = 0$, the investment should be realized as late as possible. The use of the derogation mechanism will only slightly reduce the cost of the investment itself, and the unit costs of electricity generation in the further operation of the unit will be comparable to the cost resulting from financing the investment using only the investor's own resources.

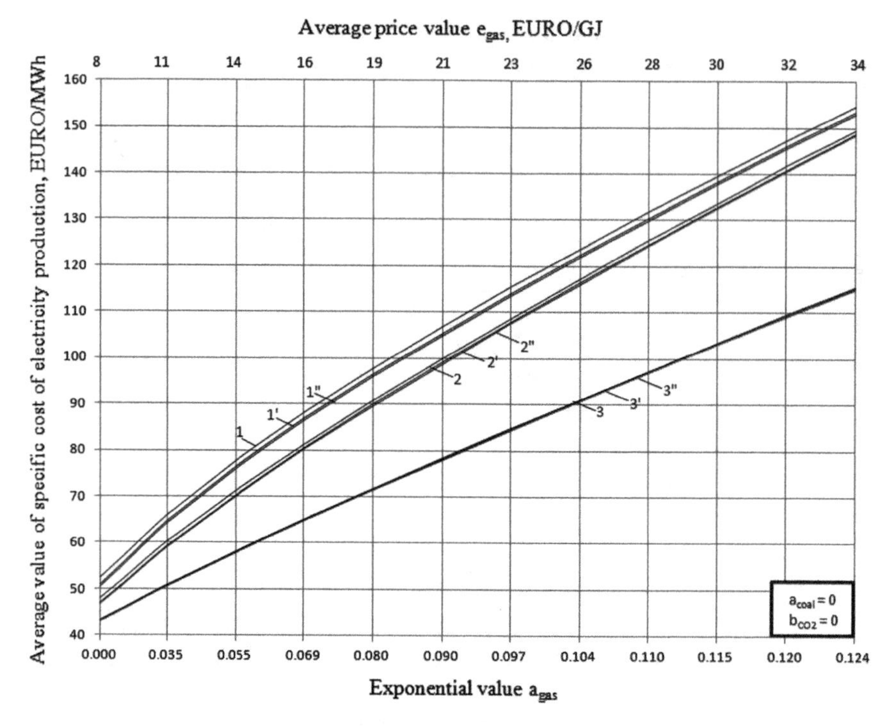

Fig. 6.10 The unit cost of electricity generation as a function of the exponent value of a_{gas} at the EUA market price = 5 € during the investment period, where: 1, 2, 3—respectively for t1 = 5 years, t1 = 10 years, t1 = 15 years, without a derogation mechanism; 1′, 2′, 3′—respectively for t1 = 5 years, t1 = 10 years, t1 = 15 years, for the derogation mechanism at the settlement price of the investment 20.38 euro/EUA (period 2015–2019); 1″, 2″, 3″—respectively for t1 = 5 years, t1 = 10 years, t1 = 15 years, for the derogation mechanism at the settlement price of the investment 14.78 euro/EUA (period 2013–2014)

As can be seen from the charts above (Figs. 6.12 and 6.13), with slightly higher EUA prices of 15 or 20 euros during the investment, it can be seen that the level of settlement of the investment against the CO_2 price is still irrelevant, because in both cases of project settlement with reference prices 14.78 or 20.38 EUR/EUA, the derogation mechanism is not much more profitable than the implementation of the investment without using the derogation mechanism, i.e. using only the investor's own resources. In each of the above cases (Figs. 6.10, 6.11, 6.12 and 6.13) the most important will be the time of launching the investment, and is more favorable for postponed modernization. Using the mechanism of derogation, with a fixed EUA prices at 5 EUR/EUA occurring on the CO_2 emission allowance market during the operation of the block, will therefore be the most advantageous for old, exhausted coal power blocks requiring renovation.

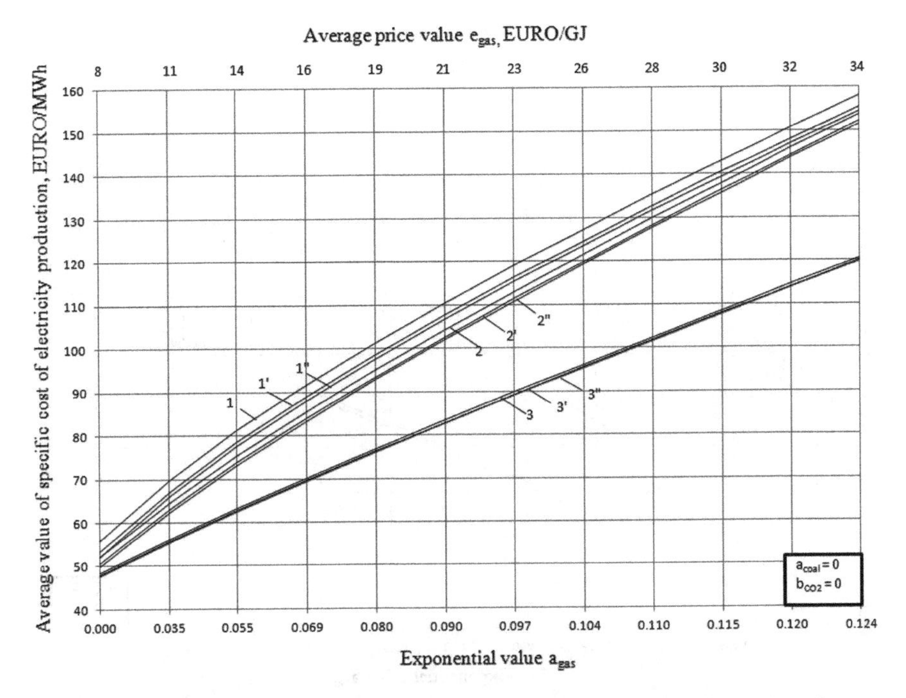

Fig. 6.11 The unit cost of electricity generation as a function of the exponent value of a_{gas} at the EUA market price = 10 € during the investment period, where: 1, 2, 3—respectively for t1 = 5 years, t1 = 10 years, t1 = 15 years, without a derogation mechanism; 1′, 2′, 3′—respectively for t1 = 5 years, t1 = 10 years, t1 = 15 years, for the derogation mechanism at the settlement price of the investment 20.38 euro/EUA (period 2015–2019); 1″, 2″, 3″—respectively for t1 = 5 years, t1 = 10 years, t1 = 15 years, for the derogation mechanism at the settlement price of the investment 14.78 euro/EUA (period 2013–2014)

6.4.3 Impact of Changes in Coal Prices on the Profitability of the Derogation Mechanism

In the next part of the study, calculations for different coal purchase prices were made (from the range of 2.7–12 EUR/GJ) during the exploitation of the unit after modernization. It was assumed that during the modernized power block operation, gas prices would remain at a constant level of 7.6 EUR/GJ, while EUA prices remained stable at 5 EUR/EUA.

The results of the calculations presented in Figs. 6.14, 6.15, 6.16 and 6.17 show that with the increase of coal prices above 5 EUR/GJ, a faster modernization of the installation is more profitable for the investor. Higher prices of EUA during modernization affect the need for faster modernization already at coal prices above 4 EUR/GJ. The use of the derogation mechanism will reduce the costs of electricity generation the most in case when the quick start of the power unit modernization

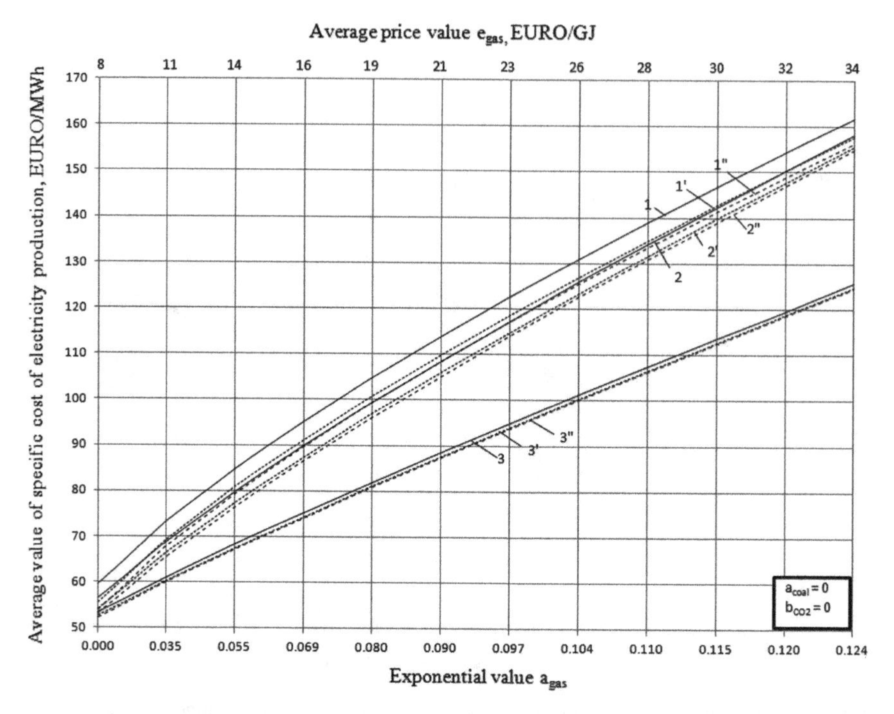

Fig. 6.12 The unit cost of electricity generation as a function of the exponent value of a_{gas} at the EUA market price = 15 € during the investment period, where: 1, 2, 3—respectively for t1 = 5 years, t1 = 10 years, t1 = 15 years, without a derogation mechanism; 1′, 2′, 3′—respectively for t1 = 5 years, t1 = 10 years, t1 = 15 years, for the derogation mechanism at the settlement price of the investment 20.38 euro/EUA (period 2015–2019); 1″, 2″, 3″—respectively for t1 = 5 years, t1 = 10 years, t1 = 15 years, for the derogation mechanism at the settlement price of the investment 14.78 euro/EUA (period 2013–2014)

investment is realized, regardless of the EUA price assumed for the allocation of free CO_2 emission allowances (reference price). This is particularly evident at the EUA price of 20 euros/EUA in force during the modernization process.

6.5 Summary and Final Conclusions

The EU ETS system is very important for increasing the cost of electricity and heat production. The mechanism of derogation introduced in it, as demonstrated in the above subsection, has a negligible impact on improving the economic efficiency of the energy sector. From 2020, there will be no more free carbon dioxide emissions, so derogations will cease to exist, and installation will have to pay for every ton of CO_2 emitted. The EU ETS system imposes in Poland a high so-called the settlement

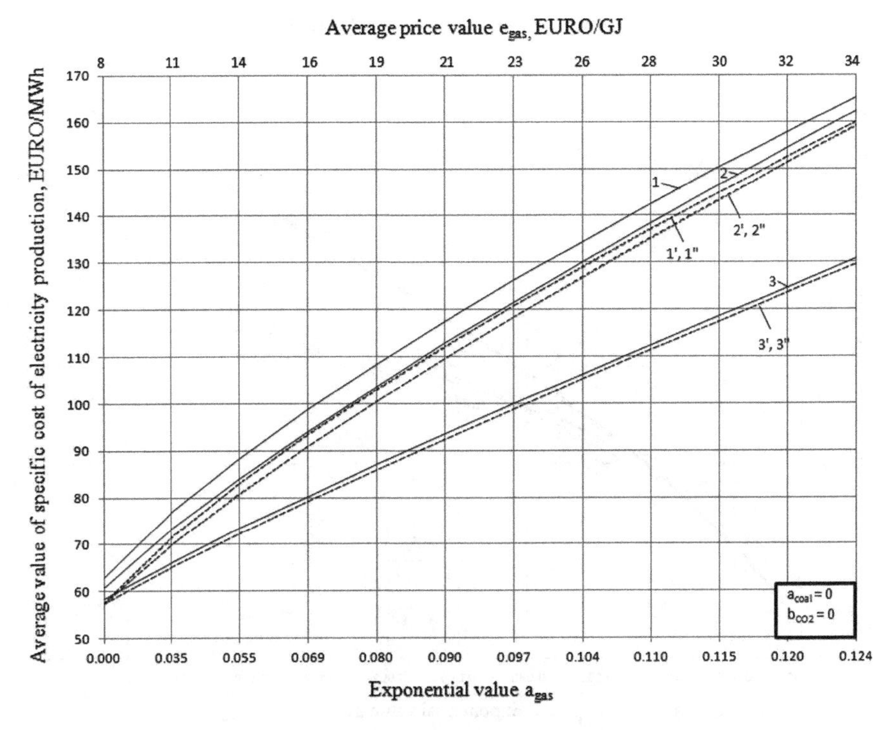

Fig. 6.13 The unit cost of electricity generation as a function of the exponent value of a_{gas} at the EUA market price = 20 € during the investment period, where: 1, 2, 3—respectively for t1 = 5 years, t1 = 10 years, t1 = 15 years, without a derogation mechanism; 1′, 2′, 3′—respectively for t1 = 5 years, t1 = 10 years, t1 = 15 years, for the derogation mechanism at the settlement price of the investment 20.38 euro/EUA (period 2015–2019); 1″, 2″, 3″—respectively for t1 = 5 years, t1 = 10 years, t1 = 15 years, for the derogation mechanism at the settlement price of the investment 14.78 euro/EUA (period 2013–2014)

(reference) price of CO_2 emissions $e_{CO_2}^{ref}$, and the higher it is, the more harmful it is to Polish electricity sector, because the lower the number of free tons of carbon dioxide (so-called EUA) allocated, which are calculated from the dependence $J_M \big/ e_{CO_2}^{ref}$ (formula 5.5).

Currently, this price is $e_{CO_2}^{ref} = 20.38$ euros per ton of emitted carbon dioxide. What is more, when the current EUA price $e_{CO_2}^{mod,\,t=t_2}$ (formula 5.5) is low, the impact of the derogation mechanism on the economic effectiveness of the modernization investment is negligible, it is of little importance to the investor. At the current value of EUA market price of only $e_{CO_2}^{mod,\,t=t_2} = 5$ EUR/MgCO2, the derogation mechanism should therefore be considered as negligible. This does not mean, however, that increasing the $e_{CO_2}^{mod,\,t=t_2}$ price is beneficial. It is the opposite. The amount of EUA tons is negligible, almost nil in comparison with tons of CO_2, for which installation owner need to buy CO_2 emission permits (formula 5.4). Increasing the $e_{CO_2}^{mod,\,t=t_2}$ price, which the European Union aims at in any way by tightening the climate and energy

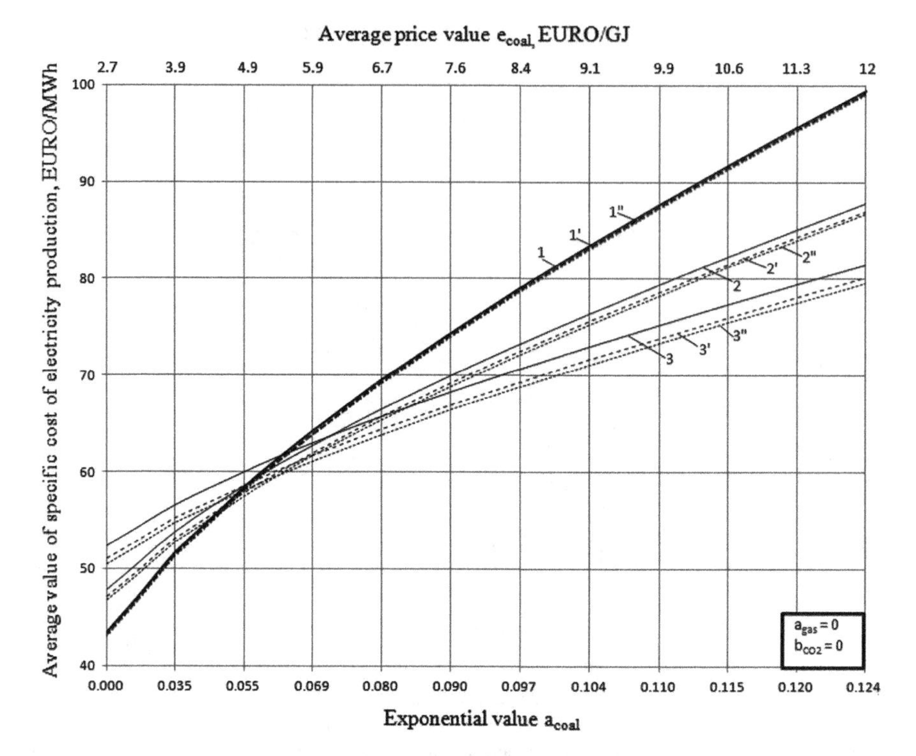

Fig. 6.14 The unit cost of electricity generation as a function of the exponent value of a_{coal} at the EUA market price = 5 € during the investment period, where: 1, 2, 3—respectively for t1 = 5 years, t1 = 10 years, t1 = 15 years, without a derogation mechanism; 1′, 2′, 3′—respectively for t1 = 5 years, t1 = 10 years, t1 = 15 years, for the derogation mechanism at the settlement price of the investment 20.38 euro/EUA (period 2015–2019); 1″, 2″, 3″—respectively for t1 = 5 years, t1 = 10 years, t1 = 15 years, for the derogation mechanism at the settlement price of the investment 14.78 euro/EUA (period 2013–2014)

policy, will result in significantly higher purchase costs of these permits, and from 2020, as already stated above, there will be no free tons of carbon dioxide emissions and installation will have to pay for each ton of CO_2 emitted. The higher the price $e_{CO_2}^{mod,\,t=t_2}$, the higher the cost of the emissions. When EUA on the market will be at the same level than the settlement price (reference value) currently amounting to 20.38 EUR/MgCO_2, Poland will pay about PLN 30 billion a year for the emissions, which is the turnkey amount of one nuclear power plant with a capacity of 1600–1700 MW.

Therefore, the only rational premise for implementing the necessary modernization investments of the Polish power industry is the revitalization of existing blocks allowing them to continue their long-term work, and a permanent increase in their power and emission reduction, rather than receiving one-time free allowances for CO_2 emissions. Otherwise, you will have to turn them off. What is more, the cost-effectiveness of energy modernization is only significantly influenced by the price

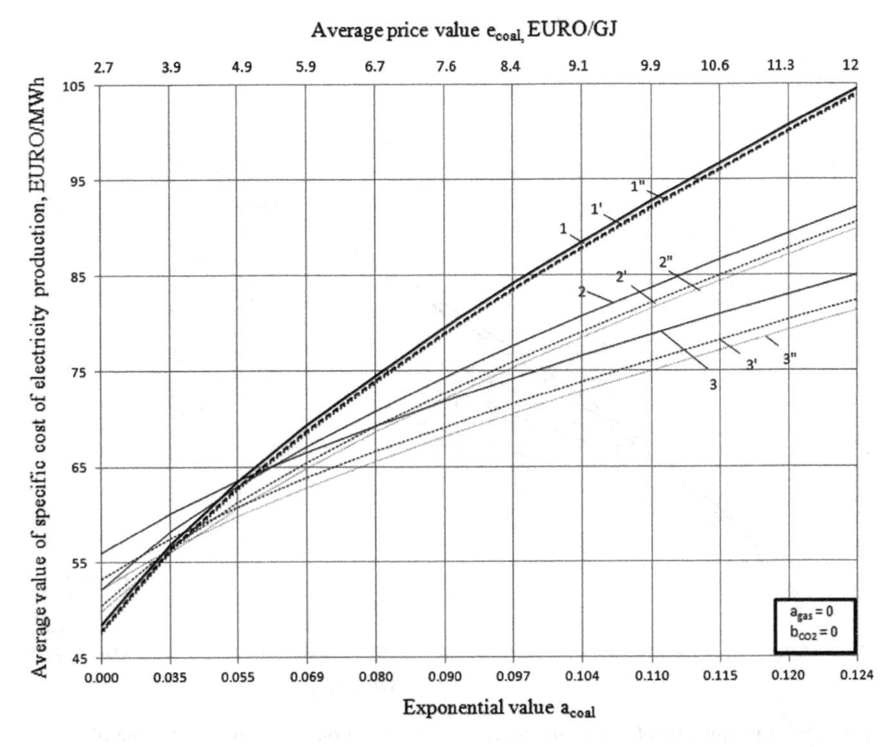

Fig. 6.15 The unit cost of electricity generation as a function of the exponent value of a_{coal} at the EUA market price = 10 € during the investment period, where: 1, 2, 3—respectively for t1 = 5 years, t1 = 10 years, t1 = 15 years, without a derogation mechanism; 1′, 2′, 3′—respectively for t1 = 5 years, t1 = 10 years, t1 = 15 years, for the derogation mechanism at the settlement price of the investment 20.38 euro/EUA (period 2015–2019); 1″, 2″, 3″—respectively for t1 = 5 years, t1 = 10 years, t1 = 15 years, for the derogation mechanism at the settlement price of the investment 14.78 euro/EUA (period 2013–2014)

relations between energy carriers, coal, gas and electricity, as demonstrated in this monograph and in the work [7].

Due to the high risk of an increase in the value of EUA prices, only investments that will contribute to a significant reduction of the emission ratio should be implemented. Moreover, the high risk associated with exceeding the duration of the modernization investment and the legal and administrative procedure for the awarding of one-off, free CO_2 emissions by EU causes reluctance to use the mechanism of derogation.

Summing up, the general conclusion can be drawn from the calculations that the reduction of the unit cost of electricity production as a result of the derogation mechanism is negligible, which almost contributes to the profitability of modernization. Necessary and sufficient conditions that must be met in order for the modernization to be economically viable are presented in monograph [3].

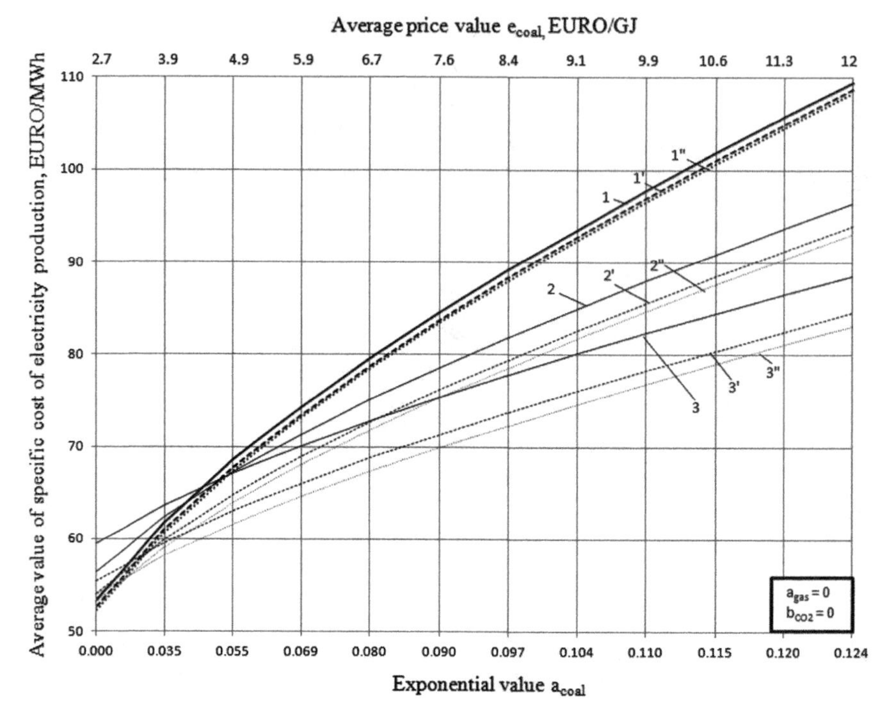

Fig. 6.16 The unit cost of electricity generation as a function of the exponent value of a_{coal} at the EUA market price = 15 € during the investment period, where: 1, 2, 3—respectively for t1 = 5 years, t1 = 10 years, t1 = 15 years, without a derogation mechanism; 1′, 2′, 3′—respectively for t1 = 5 years, t1 = 10 years, t1 = 15 years, for the derogation mechanism at the settlement price of the investment 20.38 euro/EUA (period 2015–2019); 1″, 2″, 3″—respectively for t1 = 5 years, t1 = 10 years, t1 = 15 years, for the derogation mechanism at the settlement price of the investment 14.78 euro/EUA (period 2013–2014)

Finally, it should be noted that the modernization of technologically obsolete, characterized by a low average of 30–32%, efficiency of electricity generation and, to a large extent, depreciated and in a poor technical condition of the Polish power industry, is absolutely necessary. Without it, it will be necessary to shut down Polish power plants and combined heat and power plants, and import electricity. Modernization, alongside the construction of nuclear power plants, is a rational way to develop Polish (and not only) energy sector. It should be strongly emphasized that nuclear power is currently the only technology until it is technically mastered by the thermonuclear fusion reaction that provides mankind with the supply of unlimited electricity continuously throughout the year with zero CO_2 emissions.

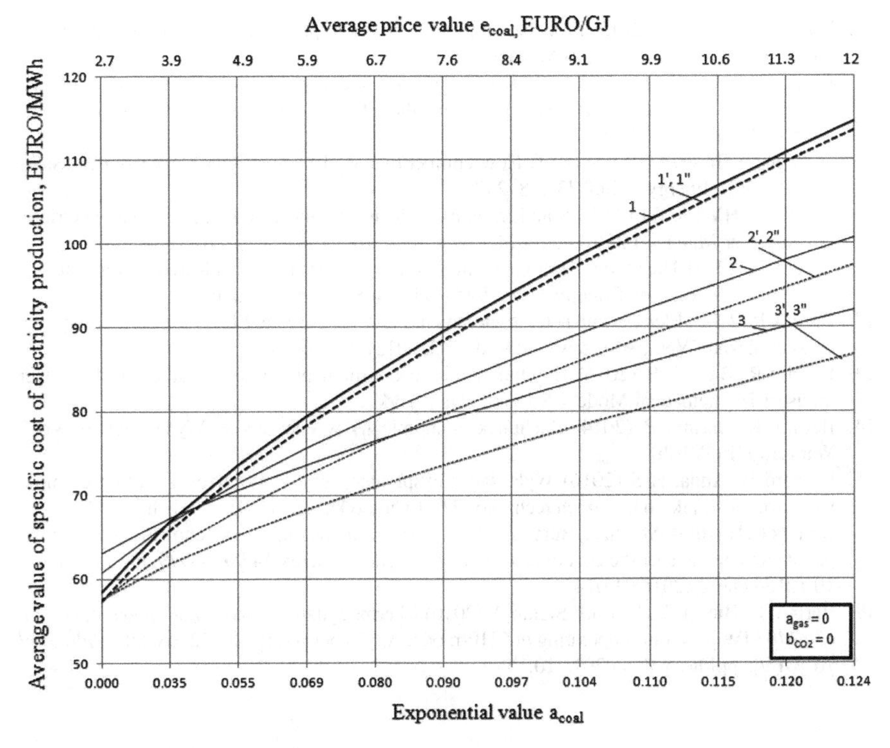

Fig. 6.17 The unit cost of electricity generation as a function of the exponent value of a_{coal} at the EUA market price = 20 € during the investment period, where: 1, 2, 3—respectively for t1 = 5 years, t1 = 10 years, t1 = 15 years, without a derogation mechanism; 1′, 2′, 3′—respectively for t1 = 5 years, t1 = 10 years, t1 = 15 years, for the derogation mechanism at the settlement price of the investment 20.38 euro/EUA (period 2015–2019); 1″, 2″, 3″—respectively for t1 = 5 years, t1 = 10 years, t1 = 15 years, for the derogation mechanism at the settlement price of the investment 14.78 euro/EUA (period 2013–2014)

References

1. Venmans Frank Maarten Jan (2016) The effect of allocation above emissions and price uncertainty on abatement investments under the EU ETS. J Clean Prod 126(10):595–606
2. Sanin ME, Violante F, Mansanet-Bataller M (2015) Understanding volatility dynamics in the EU-ETS market. Energy Policy 82:321–331
3. Bonenti F, Oggioni G, Allevi E, Marangoni G (2013) Evaluating the EU ETS impacts on profits, investments and prices of the Italian electricity market. Energy Policy 59:242–256
4. Hepburn C, Neuhoff K, Acworth W, Burtraw D, Jotzo F (2016) The economics of the EU ETS market stability reserve. J Env Econom Manage 80:1–5
5. Koch N, Fuss S, Grosjean G, Edenhofer Ottmar (2014) Causes of the EU ETS price drop: recession, CDM, renewable policies or a bit of everything?—New evidence. Energy Policy 73:676–685
6. Brink C, Vollebergh HRJ, van der Werf E (2016) Carbon pricing in the EU: Evaluation of different EU ETS reform options. Energy Policy 97:603–617

7. Perino G, Willner M (2016) Procrastinating reform: the impact of the market stability reserve on the EU ETS. J Env Econom Manage 80:37–52
8. Holt AC, Shobe WM (2016) Reprint of: Price and quantity collars for stabilizing emission allowance prices: Laboratory experiments on the EU ETS market stability reserve. J Env Econom Manage 80:69–86
9. Castagneto-Gissey Giorgio (2014) How competitive are EU electricity markets? An Assess ETS Phase II, Energy Policy 73:278–297
10. Hoffmann HV (2007) EU ETS and investment decisions: the case of the german electricity industry. Eur Manage J 25(6):464–474
11. Bartnik R (2013) The modernization potential of gas turbines in the coal-fired power industry. Thermal and Economic Effectiveness. Wydawnictwo Springer, London
12. Bartnik R (2009) Elektrownie i elektrociepłownie gazowo-parowe. Efektywność energetyczna i ekonomiczna, WNT, Warszawa 2009 (dodruk 2012) [in Polish]
13. Bartnik R, Bartnik B (2016) Hnydiuk-stefan a: optimum investment strategy in the power industry. Mathematical Models. Springer, New York
14. Bartnik R, Bartnik B (2014) Rachunek ekonomiczny w energetyce. Wydawnictwo WNT, Warszawa [in Polish]
15. Ryszard B, Anna, H-S (2016) Wyłączać z eksploatacji czy modernizować istniejące bloki węglowe? Jeśli tak, to do jakich technologii?. -Energetyka. pp. 5–14 [in Polish]
16. Bartnik R, Hnydiuk-Stefan A, Buryn Z (2018) Analysis of the impact of technical and economic parameters on the specific cost of electricity production. Energy 147:965–979. https://doi.org/10.1016/j.energy.2018.01.014
17. Bartnik R, Buryn Z, Hnydiuk-Stefan A (2016) Thermodynamic analysis of annual operation of a 370 MW power unit operating in CHP-mode. Appl Therm Eng 106:42–48. https://doi.org/10.1016/j.applthermaleng.2015.10.165

Index

A. Hnydiuk-Stefan, *Dual-Fuel Gas-Steam Power Block Analysis*, Power Systems,
https://doi.org/10.1007/978-3-030-03050-6

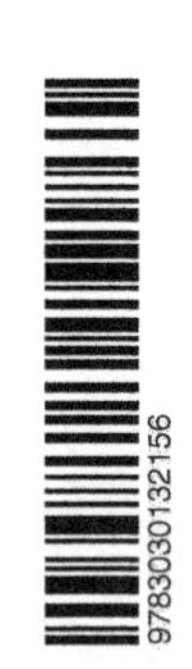
9783030132156